Atlas of *Mycobacterium tuberculosis*

Atlas of *Mycobacterium tuberculosis*

Ali Akbar Velayati and Parissa Farnia
Mycobacteriology Research Center
National Research Institute
of Tuberculosis and Lung Disease (NRITLD)
Shahid Beheshti University of Medical Sciences
Tehran, Iran

ELSEVIER

AMSTERDAM • BOSTON • HEIDELBERG • LONDON
NEW YORK • OXFORD • PARIS • SAN DIEGO
SAN FRANCISCO • SINGAPORE • SYDNEY • TOKYO

Academic Press is an imprint of Elsevier

Academic Press is an imprint of Elsevier
125 London Wall, London EC2Y 5AS, United Kingdom
525 B Street, Suite 1800, San Diego, CA 92101-4495, United States
50 Hampshire Street, 5th Floor, Cambridge, MA 02139, United States
The Boulevard, Langford Lane, Kidlington, Oxford OX5 1GB, United Kingdom

Notices
Knowledge and best practice in this field are constantly changing. As new research and experience broaden our understanding, changes in research methods, professional practices, or medical treatment may become necessary.

Practitioners and researchers may always rely on their own experience and knowledge in evaluating and using any information, methods, compounds, or experiments described herein. In using such information or methods they should be mindful of their own safety and the safety of others, including parties for whom they have a professional responsibility.

To the fullest extent of the law, neither the Publisher nor the authors, contributors, or editors, assume any liability for any injury and/or damage to persons or property as a matter of products liability, negligence or otherwise, or from any use or operation of any methods, products, instructions, or ideas contained in the material herein.

Library of Congress Cataloging-in-Publication Data
A catalog record for this book is available from the Library of Congress

British Library Cataloguing-in-Publication Data
A catalogue record for this book is available from the British Library

ISBN: 978-0-12-803808-6

Publisher: Mica Haley
Acquisition Editor: Stacy Masucci
Editorial Project Manager: Sam W. Young
Production Project Manager: Chris Wortley
Designer: Greg Harris

Typeset by TNQ Books and Journals

Dedication

This book is dedicated to
Our family
who encourage and cheer our work with affection and love.

We also dedicate this book with our deepest gratitude and warmest affection to our best scientific collaborator **Professor Gennady Konstantinovich Zhavnerko** (Belarus) who has been a constant source of knowledge and inspiration.

Contents

Preface

Atlas of Mycobacterium tuberculosis reveals the in-depth formation of *M. tuberculosis*, describes the ultrastructure of *M. tuberculosis* in the in vitro condition, and underlines how this bacterium structure relates to its function. The book was designed to combine the clarity of an atlas with the perspective of up-to date knowledge of *M. tuberculosis*. An atlas format is maintained throughout the book, with electronic or atomic micrographs on top of the pages with corresponding text below for all chapters except "Chapter 1, The Species Concept." The electronic and atomic micrographs used in this book represent the extensive works of authors performed during the years 2000–2013. However, to explain each picture, authors have used the current and primary sources of literature. For researchers interested in exploring beyond the information provided in the sections, we have included a list of references at the end of the book. Those researchers who work with *M. tuberculosis* understand very well how difficult is to work with *M. tuberculosis*. We hope this book can provide a broad foundation of knowledge in the tubercle bacilli that are considered a serious health problem throughout the world. Lastly, we would appreciate hearing from readers of our book about the ways they find it useful and the ways it could be improved.

Acknowledgments

Our special thanks go to Professor G.K. Zhavnerko (Belarus) and Mr. H. Oi Kuan (Malaysia) who helped us to assemble the original electron and atomic micrographs in 2008, 2009, 2011, and 2012. The initial collection of such micrographs led to development of this book. We are indebted to Professor J.D. Ghanavi (Iran) and Dr. P.O. Farnia (Iran) for designing and evaluating the experimental setup at different stages of cell growth. Understanding of the basic science of *Mycobacterium tuberculosis* was only possible through scientific collaboration with scientists in this field. Therefore we greatly appreciate the sincere support and suggestions of the following: Professor M. Bose (India), Professor F. Mohammadi (Iran), Professor M.R. Masjedi (Iran), Professor V. Vincent (France), Professor A.Z. Zarifi (Iran), Dr. C. Gutierrez (France), Dr. P. Tabarsi (Iran), Dr. M. Marjani (Iran), Dr. A. Moniri (Iran), Dr. M. Pourabdollah (Iran), Dr. M. Varahram (Iran), Dr. A.M. Farahbod (Iran), Professor M. Bahadori (Iran), Professor S.J. Tabatabaei (Iran), Dr. E. Mortaz (Iran), Dr. H. Jamaati (Iran), Dr. M. Kazempour Dizaji (Iran), Dr. A. Bahrami (Iran), Professor T. Azmi Ibrahim (Malaysia), Mr. R.Z. Haroun (Malaysia), Professor N.N. Poleschuyk (Belarus), Professor L. Ttitov (Belarus), Professor G. Fada (Italy), Professor M. Ridell (Sweden), Professor L. Olof Larsson (Sweden), Professor D.M. Cirillo (Italy), Dr. L. Rubanik (Belarus), Dr. M. Arjomandzadegan (Iran), Dr. A. Astashonok (Belarus), Miss. R. Abdi (Iran), Dr. M. Mirsaeidi (United States), Dr. E.M. Ashlee (United States), Professor G.H. Cassell (United States), and Professor S. Hoffner (Sweden). We also thank the laboratory staff at the Mycobacteriology Research Center (Iran).

Chapter 1

The Species Concept

Mycobacteria represent a very ancient genus of bacteria (Bates and Stead, 1993; Benedek, 2004; Donoghue et al., 2004; Taylor et al., 2005). They have been around for millions of years and have been adapted to nearly every environment on the Earth (ie, water, soil, dust, and air; Falkinham, 1996, 2002, 2009, 2010; Whiley et al., 2012; Collins et al., 1984; Nassal et al., 1974; Papapetropoulou et al., 1997; Hruska and Kaevska, 2012; Ghodbane et al., 2014). Their antiquity and their great age were supported by valuable evidence collected from the study of mummified ibises and cats from Egyptian tombs (Konomi et al., 2002; Daniel, 2006). The genus *Mycobacterium* encompasses several medically important species that cause chronic pulmonary and nonpulmonary diseases (Barksdale and Kim, 1977; Shinnick and Good, 1994; Hartmans and De Bont, 2006; Saviola and Bishai, 2006). The generic name of *Mycobacterium* was initially designated to the way that these organisms grew in a mold-like fashion on liquid culture media (Lehmann and Neumann, 1896). The main characteristic of mycobacteria are the absence of motility, the morphology of the bacilli (slightly curved and rod shaped), and the resistance to acid-alcohol after coloration with phenicated fuchsin (Ziehl-Neelsen stain; Ehrlich, 1887; Koch, 1882; Goodfellow and Wayne, 1982; Wayne and Kubica, 1986; Rastogi et al., 2001; Figs. 1.1 and 1.2).

They belong to the family Mycobacteriaceae (Table 1.1) within the order Actinomycetales (Stackebrandt et al., 1997). Actinomycetales include diverse microorganisms in ecological and morphological terms (Goodfellow and Wayne, 1982; Wayne and Kubica, 1986). The genus *Mycobacterium* is considered to be closely related to other members of Actinomycetales (ie, *Corynebacterium*, *Nocardia,* and *Rhodococcus*; Lechevalier and Lechevalier, 1970; Collins et al., 1988; Vincent, 1994; Table 1.2).

Differentiation between the *Mycobacterium* and allied taxa can be easily performed by the length of the carbon backbone, the number of unsaturated links, the presence of supplementary oxygenated function, and the esters produced on pyrolysis (Lévy-Frébault and Portaels, 1992; Vincent, 1994; Rastogi et al., 2001). However, the Ziehl-Neelsen stain for acid fastness remains the most obvious method to quickly identify *Mycobacterium* (Barksdale and Kim, 1977). In general, the color and morphology of *Mycobacterium* grown on solid culture media are the main features of these

Atlas of *Mycobacterium tuberculosis*.

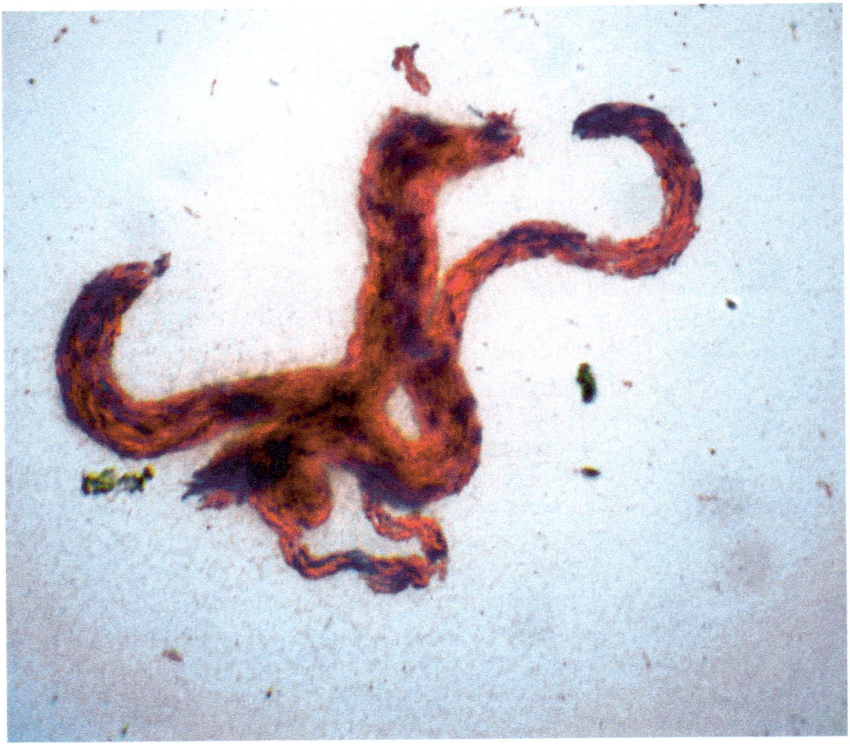

FIGURE 1.1 Visualization of *M. tuberculosis* using the Ziehl-Neelsen stain. The most important characteristic of *Mycobacterium* is their acid fastness, was discovered by Ehrlich in 1882. Organisms may be stained with any of a variety of basic dyes, but they cannot be subsequently decolorized with dilute acids. Their acid fastness is due to the presence of unsaponifiable wax of the nature of hydroxy acid known as mycolic acid. Ziehl-Neelsen staining was first described by two German doctors—the bacteriologist Franz Ziehl (1859–1926) and the pathologist Friedrich Neelsen (1854–1898).

microorganisms (Tárnok and Tárnok, 1970; David, 1974; McCready and Ratledge, 1978). Most species are whitish or cream-colored colonies (Fig. 1.3), but especially among the rapid growers there are also many bright yellow (Fig. 1.4) or orange species containing carotenoid pigments (Fig. 1.5). The possession of such pigments and the ability of strains to produce them either in the dark (scotochromogenic species) or in response to light (photochromogenic species) have been used to classify some potentially pathogenic *Mycobacterium* (Juhlin, 1967).

The morphology of mycobacterial colonies on solid culture media is a stable character of a given strain, although variants often arise by spontaneous mutations (Fregnan and Smith, 1962; Vestal and Kubica, 1966). The basic types of colonies, are "rough" and "smooth" (Figs. 1.6 and 1.7).

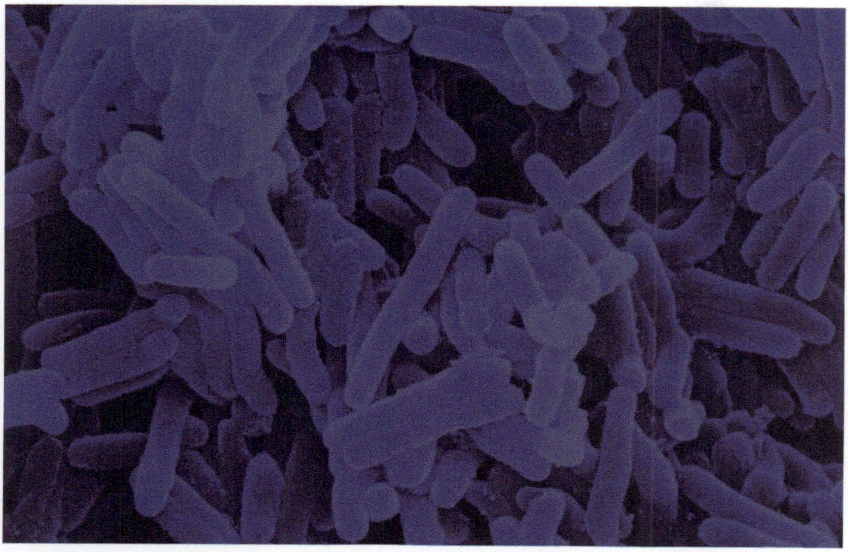

FIGURE 1.2 Transmission electron microscopy (TEM) of the type-species of *M. tuberculosis*. Species was first seen by Koch in 1882. It was in the form of a bacillus, 1–4 μm long and 0.3–0.6 μm in diameter (Runyon et al., 1974).

CLASSIFICATION OF *MYCOBACTERIUM*

The identification of the genus *Mycobacterium* began in the late 1800s with the discovery of the leprosy bacillus (originally named *Bacillus leprae*; Hansen, 1880) and the tubercle bacillus (originally named *Bacterium tuberculosis*; Zopf, 1883). However, their classification began in 1896 when Lehmann and Neumann placed the genus *Mycobacterium* in the Mycobacteriaceae family, Actinomycetales order, and Actinomycetes class. In the early period classification was based on one or more characters or patterns of characters that all members of a group have and that members of other groups do not have. The defined characteristics were morphology (rod shaped, nonmotile), and acid-alcohol fastness, which led to difficulties in clearly distinguishing members of the genus *Mycobacterium* from closely related genera (ie, *Nocardia*, *Rhodococcus*, and *Corynebacterium*). Therefore additional criteria were developed for differentiating these genera, including gram staining, growth rate, and mycolic acid profiles (Castelnuovo et al., 1968; Kwapinski et al., 1970; Lechevalier et al., 1971; Ridell and Norlin, 1973). Runyon (1959) classified members of mycobacteria on the basis of their growth, pigmentation, and clinical significance (Table 1.3).

Groups I, II, and III included only slowly growing mycobacteria (ie, organisms that require more than 1 week to grow) whereas Group IV included rapidly growing mycobacteria that require 1 week or less for culture. Runyon Group I includes photochromogenic strains, in which colonies

TABLE 1.1 Classification of *Mycobacterium*

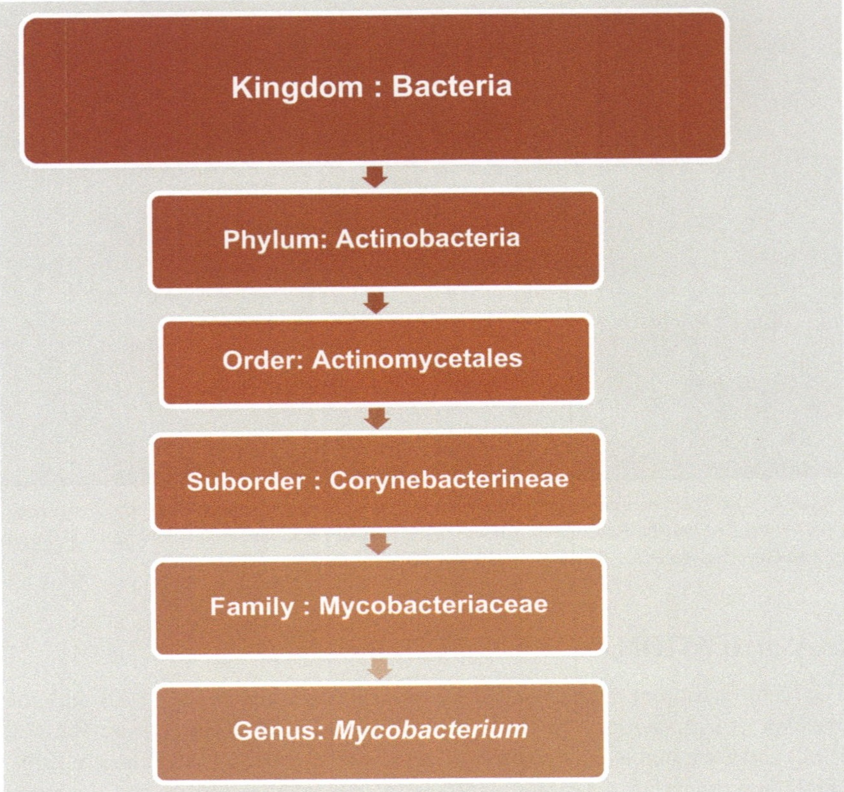

acquire pigmentation only in the presence of light. Group II includes scotochromogenic strains that are pigmented in the presence or absence of light. Group III comprises nonchromogenic strains (Runyon and Selin, 1959). This classification system is very useful; however, it has some limitations. For example, members of the *Mycobacterium avium-intracellulare* complex are nonphotochromogens in Runyon's scheme, but some isolates produce slightly pigmented colonies, potentially causing incorrect classification as a scotochromogen. Moreover, it has been repeatedly observed that particular slow-growing strains are scored as giving a negative response but give positive responses upon longer incubation (Sneath, 1970). With the identification of more characters in *Mycobacterium*, investigators began using an Adansonian classification scheme known as numerical taxonomy. The principles of

TABLE 1.2 Chemical Characteristics of *Mycobacterium* and Closely Related Members

Genus	GC Content of DNA (mol %)	Mycolic Acids		*N*-glycolyl in Glycan Moiety of Cell Wall
		Overall Size (Number of Carbons)	Ester Pyrolysis Products (Number of Carbons in Chain)	
Nocardia	64–69	46–60	12–18	+
Mycobacterium	62–70	60–90	22–26	+
Cordona	63–73	48–66	16–18	–
Corynebacterium	48–59	22–38	8–18	–
Tsukamurella	63–73	64–78	20–22	+
Rhodococcus	63–73	34–52	12–18	+

Goodfellow and Wayne (1982), Vincent (1994), Hartmans et al. (2006), Rastogi et al. (2001).

Adansonian taxonomy are (1) it is based upon as many characters as possible, (2) each character is given equal weight, (3) the overall or phonetic similarity between each pair of strains is calculated from the number of characters that they have in common, and (4) organisms that share many characteristics are grouped together (Bojalil et al., 1962; Sneath, 1970; Bradley and Bond, 1974). In this system it is essential that many organisms be examined with respect to many characters. As expected, the discriminatory power of the approach as well as the reliability and significance of the clustering of strains increases as the number of independent characteristics measured increases. In general, it provided a thorough description of each species and facilitated the discovery of tests that could distinguish reliably in various species (Shinnick and Good, 1994). However, when the numbers of strains and properties studies increased, the assimilation of a large amount of phonetic information into a coherent picture of relatedness became more difficult. To facilitate the analysis of relatedness of strains and species, an International Working Group on Mycobacterial Taxonomy (IWGMT) was established (1967). Through the concerted efforts of the IWGMT and individual investigators, the *Mycobacterium* genus has become one of the best described genera. Wayne et al. (1971) performed a cooperative numerical taxonomic analysis of slowly growing scotochromogenic mycobacteria of Runyon's Group II.

FIGURE 1.3 The colonies of *Mycobacterium* on solid culture media can be of various shades from cream to yellow and deep orange color. *M. tuberculosis* produces cream-colored buff, rough colonies after 2–3 weeks inoculation on Löwenstein–Jensen medium.

These workers defined four clusters; three of the clusters corresponded to named species (*Mycobacterium flavescens*, *Mycobacterium gordonae*, and *Mycobacterium scrofulaceum*). The fourth cluster may represent a new species. Cross and Goodfellow (1973) proposed a similar but more extensive grouping; he separated *Mycobacterium leprae* from *Mycobacterium* genus. On the basis of pigment production he subdivided the rapid grower *Mycobacterium* into nonpigmented mycobacteria and scotochromogenic mycobacteria (Table 1.4). This division had no formal taxonomic standing but was clinically useful in identification schemes.

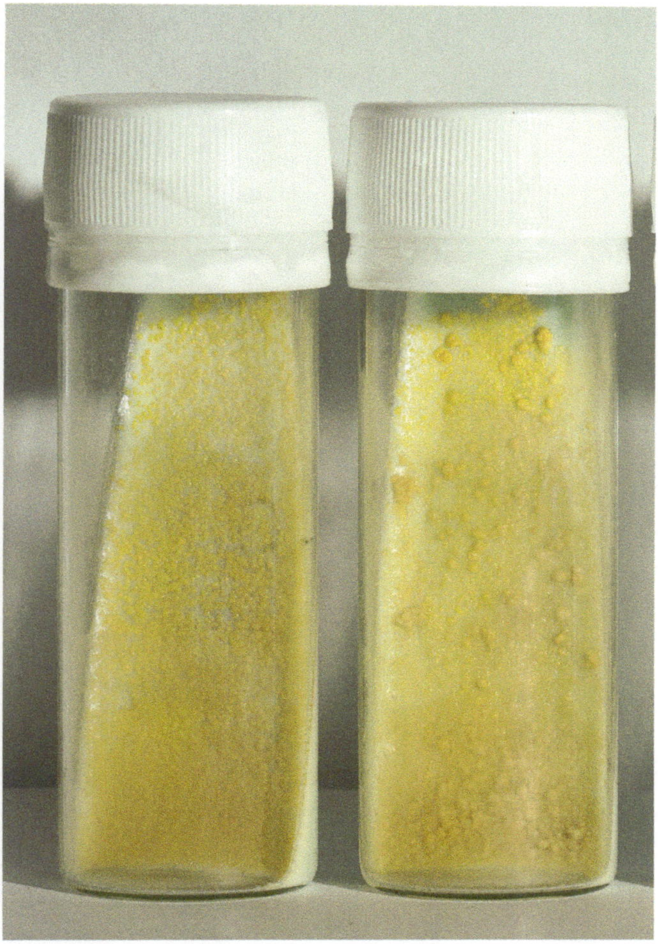

FIGURE 1.4 The colonies of photochromogenic strains of *Mycobacterium* (Runyon classification), when exposed briefly to light, become brilliant yellow.

Kubica (1973) has formulated a simplified taxonomic scheme that facilitates identification of most clinically important mycobacteria recovered from man. The scheme is based upon growth rate and 11 other simple tests selected from 82 taxonomic characters (Bradley and Bond, 1974). The selected tests for the abbreviated scheme were niacin production, nitrate reduction, catalase production, photochromogenicity, tween 80 hydrolysis, tellurite reduction, sodium chloride tolerance; aryl sulfatase activity, and growth on MacConkey medium (Table 1.5).

Bradley and Bond (1974) described some of the basic principles of measuring DNA homology and its significance and application in mycobacterial

FIGURE 1.5 The colonies of scotochromogenic strains of *Mycobacterium* (Runyon classification). The grown colonies in the presence or absence of light are pigmented (from dark yellow to deep orange).

classification. The DNA hybridization experiments showed a low degree of homology between mycobacterial species (Baess, 1974; Bradley, 1973). Wayne (1978) reviewed some of the limitations of applying conventional numerical taxonomy to the genus and discussed the importance of evaluating numerical classifications in respect to relationships based on other criteria. An objective taxonomical classification may be obtained by correlating the results of numerical taxonomic studies with the determination of DNA base compositions and measurements of DNA homology (Grange, 1982; Imaeda et al., 1982). In 1980 the International Code of Nomenclature of the bacteria (ICNB) published the name of 41 *Mycobacterium* species (Table 1.6).

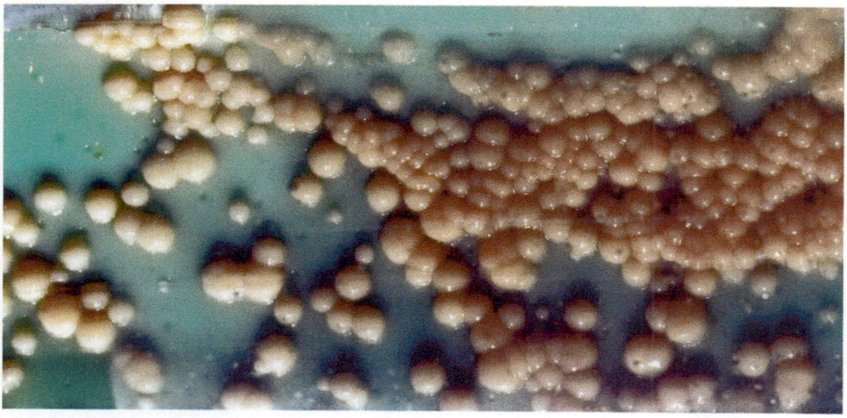

FIGURE 1.6 Smooth colonies of *Mycobacterium* grown on Löwenstein–Jensen medium.

FIGURE 1.7 Rough colonies of *Mycobacterium* grown on Löwenstein–Jensen medium.

TABLE 1.3 Runyon's Classification Schemes of the Nontuberculous Mycobacteria

Groups	Classification	Description of Colonies
Group I	Photochromogen	Not pigmented unless exposed to light
Group II	Scotochromogen	Pigmented when grown in the dark and in the light
Group III	Nonphotochromogen	Not pigmented when grown in the dark or in the light
Group IV	Rapid grower	Growth on solid media in ≤7 days

TABLE 1.4 Cross and Goodfellow System of Classification (1973)

Group	Name of Species
M. Leprae and other noncultivable species (eg, *M. lepraemurium*)	
M. tuberculosis (also include *M. bovis*, *M. microti*)	
Photochromogenic mycobacteria	*M. kansasii, M. marinum*
Slowly growing scotochromogenic mycobacteria	*M. scrofulaceum, M. marianum, M. gordonae, M. flavescens*
Nonchromogenic slow-growing niacin-negative mycobacteria	*M. avium, M. intracellulare, M. terrae, M. gastri, M. ulcerans*
Rapidly growing nonpigmented mycobacteria	*M. fortuitum, M. abscessus, M. smegmatis, M. diernhoferi, M. chitae*
Rapidly growing scotochromogenic mycobacteria	*M. gilvum, M. obuense, M. thermoresistibile, M. vaccae, M. duvalii, M. flavescens, M. phlei, M. rhodesiae*

TABLE 1.5 Kubica's Taxonomic Scheme of Mycobacteria

Group I	*M. kansasii, M. marinum*
Group II	*M. scrofulaceum, M. gordonae, M. flavescens*
Group III	*M. xenopi, M. avium, M. intracellulare, M. gastri, M. terrae, M. trivale*
Group IV	*M. fortuitum, M. vaccae, M. abscessus, M. smegmatis, M. phlei*
Principal human pathogens	*M. tuberculosis, M. bovis, M. africanum*

Goodfellow and Wayne (1982) have updated the description of the genus in their review:

> *[M]ycobacteria are aerobic, acid-alcohol fast actinomycetes that usually produce slightly curved or straight rods and coccoid elements; aerial hyphae are usually absent. Strains have meso-diamino-pimelic acid, arabinose and* galactose *in the peptidoglycan, produce straight-chain unsaturated fatty acids, contain complex mixtures of mycolic acids (C-60 to C-90) and dihydrogenated menaquinones with nine isoprene units, and the G + C content of their DNA ranges from 62% to 70%.*

TABLE 1.6 **The List of Mycobacteria Published by International Code of Nomenclature of Bacteria**

Taxon	Growth Rate	Taxon	Growth Rate
M. africanum	Rapid	*M. malmoense*	Slow
M. asiaticum	Slow	*M. marinum*	Slow
M. aurum	Rapid	*M. microti*	Slow
M. avium	Slow	*M. nonchromogenicum*	Slow
M. bovis	Slow	*M. neoaurum*	Rapid
M. chelonae	Rapid	*M. parafortuitum*	Rapid
M. chitae	Rapid	*M. paratuberculosis*	Slow
M. duvalii	Rapid	*M. phlei*	Rapid
M. farcinogenes	Rapid	*M. scrofulaceum*	Slow
M. flavescens	Rapid	*M. senegalense*	Rapid
M. fortuitum	Rapid	*M. simiae*	Slow
M. gadium	Rapid	*M. smegmatis*	Rapid
M. gastri	Slow	*M. szulgai*	Slow
M. gilvum	Rapid	*M. terrae*	Slow
M. gordonae	Slow	*M. thermoresistibile*	Rapid
M. haemophilum	Slow	*M. triviale*	Slow
M. intracellulare	Slow	*M .tuberculosis*	Slow
M. kansasii	Slow	*M. ulcerans*	Slow
M. komossense	Rapid	*M. vaccae*	Rapid
M. leprae	Slow	*M. xenopi*	Slow
M. lepraemurium	*Slow*		

Kazda (1983) classified the mycobacteria into four groups on the basis of ecologically relevant properties (Table 1.7).

In 1984 Kubica and Wayne included three new species of *Mycobacterium* (*M. chelonae* subsp. *chelonae*, *M. chelonae* subsp. *abscessus*, *M. sphagni*), which brings the number of species to 44. By the end of 1983 there were 52 described species. In early 1999 several studies were published that covered the phylogenetic structure of the fast- and slow-growing mycobacteria (Smida et al., 1988;

TABLE 1.7 On the Basis of Ecological Properties, Kazda (1983) Classified Mycobacteria Into Four Groups

Classification	Species
Obligate pathogens (unable to multiply outside living beings)	*M. tuberculosis*
	M. bovis
	M. africanum
	M. asiaticum
	M. malmoense
	M. microti
	M. simiae
	M. szulgai
	M. haemophilum
Facultative pathogens	*M. leprae*
	M. paratuberculosis
	M. ulcerans
Potential pathogens	*M. avium*
	M. chelonae
	M. fortuitum
	M. intracellulare
	M. kansasii
	M. marinum
	M. senegalense
	M. scrofulaceum
	M. xenopi
Saprophytic species	*M. gordonae*
	M. nonchromogenicum
	M. triviale
	M. terrae
	M. gastri
	Most of the rapid-grower species

Rogall et al., 1990; Stahl and Urbance, 1990; Pitulle et al., 1992). These data by and large supported the clustering of mycobacteria according to their growth behavior and indicated that the mostly clinical, slow-growing strains evolved from their fast-growing relatives (Pitulle et al., 1992). The analysis of large information-bearing molecules such as nucleic acids and nucleotide sequences of ribosomal RNA (rRNA) genes have been used to construct phylogenetic trees among mycobacteria (Wayne and Diaz, 1985; Stackebrandt et al., 1987). The basic ideas that make rRNA genes particularly useful as molecular evolutionary clocks are (1) rRNA genes are ubiquitous among the bacteria and eukaryotes, (2) rRNA functions have been highly conserved over extremely long evolutionary times, (3) rRNA sequences are sufficiently long to contain a large amount of phylogenetic information, and (4) rRNA molecules display extensive secondary structure that helps align the nucleotide sequences and identify differences (Shinnick and Good,1994). The phylogenetic trees of the genus *Mycobacterium* using a sequence of 16S rRNA genes showed that all species of *Mycobacterium* are closely related to each other and distantly related to members of other genera (Pitulle et al., 1992; Rogall et al., 1990; Shinnick and Good, 1994).

At present more than 150 recognized or proposed species meet the standard for inclusion in the genus of *Mycobacterium*. These species are usually grouped into two major divisions—rapidly growing and slowly growing—on the basis of the time required for visible colonies to appear on a solid medium after plating sufficiently dilute suspensions to give well-separated colonies (ie, presumably colonies arising from single cells). The appearance of colonies requires fewer than 7 days for the rapidly growing species and more than 7 days for the slowly growing species (Table 1.8).

MYCOBACTERIUM TUBERCULOSIS COMPLEX

Mycobacterium tuberculosis complex (MTC) constitutes a remarkably genetically homogeneous group. They are characterized by 99.9% similarity at the nucleotide level and identical 16S rRNA sequences (Böddinghaus et al., 1990; Sreevatsan et al., 1997), but they differ widely in terms of their host tropisms, phenotypes, and pathogenicity (Brosch et al., 2002; Rastogi and Sola,2007; Wirth et al., 2008). The MTC includes *Mycobacterium tuberculosis*, *Mycobacterium africanum*, *Mycobacterium bovis*, *Mycobacterium microti*, *Mycobacterium canettii*, *Mycobacterium caprae*, *Mycobacterium pinnipedii*, *Mycobacterium suricattae*, *Mycobacterium mungi*, *Mycobacterium dassie*, and *Mycobacterium oryx*. From those species, *M. tuberculosis*, "a pathogen of the mammalian respiratory system," is the most well-known member, infecting more than one-third of world's human population; it is also able to infect animals that have contact with humans (Borgdorff, 2000; Smith, 2004; Dye, 2006; Montoro and Rodriguez, 2007; van Soolingen et al., 2007).

M. bovis displays the broadest spectrum of host infection, affecting humans, domestic or wild bovines, and goats (Karlson and Lessel, 1970; Grange et al., 1996; Phillips et al., 2003; O'Brien, 2011). *M. bovis* is usually transmitted to

TABLE 1.8 Routine Tests Used for Classification of *Mycobacterium*

Slowly Growing Species	Rapidly Growing Species
Growth at different temperatures (25, 30, 37, 42, and 45°C)	Growth at different temperatures (25, 30, 37, 42, and 45°C)
Pigmentation	Pigmentation
Biochemical tests	Biochemical tests
Heat-resistant catalase	Niacin production
Semiquantitative catalase	Nitrate reductase
Tween 80 hydrolysis	Iron uptake
Urease	Acid phosphatase
Niacin production	3-day arylsulfatase
Nitrate reductase	10-day arylsulfatase
Acid phosphatase	Pyrazinamidase
3-day arylsulfatase	α-Esterase
10-day arylsulfatase	β-Esterase
Pyrazinamidase	β-Galactosidase
α-Esterase	Degradation of *p*-aminosalicylate
β-Esterase	Acetamidase
β-Galactosidase	Isonicotinamidase
Mycolic acid profile	Allantoinamidase
Mycolic acid pyrolysis esters	Succinamidase
DNA–DNA hybridization analysis	Production of acid from L-arabinose, xylose, or dulcitol
16S ribosomal RNA sequence	Utilization of xylose, oxalate, citrate, sorbitol, mannitol, or trehalose
	Mycolic acid profile
	Mycolic acid pyrolysis esters
	DNA–DNA hybridization analysis
	16S ribosomal RNA sequence

humans by infected milk, although it also spread via aerosol droplets (Velayati et al., 2007). *M. bovis* BCG, historically known as Calmette et Guérin, is prepared from an attenuated strain of the live bovine tuberculosis bacillus *M. bovis* (King and Park, 1929; Aronson, 1948; Andersen, 1948). *M. bovis* BCG is the only vaccine used in tuberculosis prevention during early childhood (Murohasi, 1948; Price, 1949; Trunz et al., 2006; WHO, 2007; Roy et al., 2014). *M. canettii* and *M. africanum* are closely related to *M. tuberculosis*. *M. canettii* was first reported in 1969 (Canetti), has shorter generation time than clinical isolates of *M. tuberculosis*, and presented a unique characteristic phenolic glycolipid and lipooligosaccharide (van Soolingen et al., 1997). Tuberculosis caused by both *M. canettii* and *M. africanum* appears to be an emerging disease in African patients or those of African ancestry (Forrellad et al., 2013). *M. microti* is a rodent pathogen, usually isolated from voles, and it can also cause diseases in immunocompromised human patients (Brosch et al., 2002). *M. caprae* has been recognized mainly in central Europe, where it has been occasionally isolated from tuberculous lesions from

TABLE 1.9 Spoligotyping Subdivided *M. tuberculosis* Into Seven Main Lineages

The main lineages within the *M. tuberculosis* species
1. The East-African-Indian (EAI) lineage
2. The Beijing lineage
3. The Central-Asian (CAS) or Delhi lineage
4. The Haarlem family
5. The Latin American and Mediterranean (LAM) family
6. The X family: The European IS6110 low banders
7. The T families and others

cattle, pigs, red deer (*Cervus elaphus*), and wild boars (*Sus scrofa*). Its isolation from humans has also been described (García-Rodríguez et al., 2011). *M. pinnipedii*, *M. suricattae*, *M. mungi*, *M. dassie*, and *M. oryx* can infect seals, meerkats, mongoose, dassie, and the Bovidae family (Wagner and Bokkenheuser, 1961; Van Ingen et al., 2012; Alexander et al., 2010).

Today, based on genomic analysis, it has been suggested that members of the MTC evolved from a common ancestor via successive DNA deletions/insertions resulting in the present speciation of *Mycobacterium* (Forrellad et al., 2013). However, the advent of molecular methods and the discovery of the polymorphic nature of the direct repeat (DR) locus (Groenen et al., 1993; Kamerbeek et al., 1997) with subsequent development of the spoligotyping method based on DR locus variability introduce more modern concepts and tools for understanding MTC genotyping (Sola, 1999; Table 1.9). Later, the discovery of tandem repeat loci in the MTC (Frothingham and Meeker-O'Connell, 1998; Supply et al., 2001; Baker et al., 2004), systematic single-nucleotide polymorphism genotyping (Filliol et al., 2006; Gutacker et al., 2006), and large sequence polymorphism (Mostowy and Behr, 2002; Tsolaki et al., 2005; Rastogi and Sola, 2007) was introduced. These molecular approaches favor of the existence of lateral genetic transfer in the precursor of the MTC and allowed the definition of genetically different lineages among *M. tuberculosis* species.

The East African-Indian lineage was first described in Guinea-Bissau (Källenius et al., 1999) and was shown to be frequent in Southeast Asia, India, and East Africa (Kremer et al., 1999). This group of strains is characterized by a low number of IS6110 copies. A subgroup of these strains harboring a single copy of IS6110 was shown to be widespread in Malaysia, Tanzania, and Oman (Fomukong et al., 1994). These lineages are characterized by the absence of spacers 29–32, the presence of spacer 33, and the absence of spacer 34. This lineage was shown to belong to Cluster Group 1 or Cluster I (Filliol et al., 2006; Gutacker et al., 2006).

The Beijing genotype belongs to the principal genetic Group 1 of Sreevatsan, and its specific spoligotype pattern is the absence of spacer 1–33 and the presence of spacer 34–43 that was discovered in 1995 (van Soolingen et al., 1995).

The Central-Asian (CAS) or Delhi lineage with its specific spoligotype pattern that is absence of spacers 4–27 and 23–34 is shown to be endemic in Sudan, other sub-Saharan countries, and Pakistan (Brudey et al., 2006). This spoligotype-signature shows numerous variants and several subgroups such as CAS1-Kili (for Kilimanjaro) and CAS1-Dar (for Dar-es-Salaam), which have already been defined on the basis of new spoligotype-signatures that are specific for each new clonal complex (McHugh et al., 2005; Eldholm et al., 2006).

The Haarlem family was described in the Netherlands in 1999 (Kremer et al., 1999). The spoligotype is characterized by the absence of spacer 31, which is due to the presence of a second copy of IS6110 in the DR region (Groenen et al., 1993). Three main spoligotype-signatures define the variants H1 to H3 (Filliol et al., 2002). The LAM family is frequent in Mediterranean countries, and its presence in Latin America (Zumárraga et al. 2014) is supposed to be linked to the Lusitanian-Hispanian colonization of the New World. The family is characterized by absence of spacers 21–24 in the spoligotyping (Sola et al., 2001). The X family of strains is defined by two concomitant features: a low number of IS6110 copies and the absence of spacer 18 in the spoligotyping (Sebban et al., 2002). The X family was also the first group identified in Guadeloupe (Sola et al., 1997) and French Polynesia (Torrea et al., 1995). Specific epidemic variants of this genotype family were described in South Africa (Streicher et al., 2004). The T families are strains that miss spacers 33–36 and can hardly be classified in other groups.

Chapter 2

Microscopic Anatomy of *Mycobacterium tuberculosis*

To a light microscopist the most important characteristic of mycobacteria is their acid fastness, discovered by Ehrlich in 1882. This acid fastness is of immense practical importance in rapidly identifying mycobacteria, but this examination cannot resolve the internal details of mycobacteria, with the exception of the "banding" pattern caused by lipid droplets (Nassau and Hamilton, 1957; Schaefer and Lewis, 1965) that is seen in some species (Harada, 1976; Draper, 1982). These droplets appear as unstained spherules at regular intervals inside of the bacilli and might represent the deposits of lipophilic material that have a storage function (Garton et al., 2002). However, because accurate judgment of the internal organelles of mycobacteria with light microscopy is technically impossible, investigators used more sophisticated microscopes to reveal the detailed information. In fact, the work in electron microscopy of mycobacteria started nearly 50 years ago (Rosenblatt et al., 1942), and in some cases it has become possible to correlate physical structure with chemical constituents of the bacterial cell. Historically, in stained whole cells, three types of internal structures were identified:

1. lipid vacuoles (Arima et al., 1972),
2. dense bodies usually near the poles of the cell (Glauert and Brieger, 1995), and
3. a structure originally considered to be analogous to mitochondria (Mudd and Wintersheid, 1953).

The dense bodies are believed to consist of polyphosphate, which may form an energy store in the cell, analogous to the β-hydroxybutyrate granules common in other types of bacteria. The third type of structure, sometimes also containing polyphosphate, was apparently also the site of oxidation–reduction reactions (Ariji et al., 1968). It is now known to correspond to the mesosome-like body (Fig. 2.1).

Developments in analytical techniques combined with knowledge of chemistry of the mycobacterial envelope have resulted in a thorough understanding of mycobacterial cell wall and its lipids (Draper, 1982; Brennan, 1994; Ortalo-Magné et al., 1996; Daffé and Darper, 1998; Draper and Daffé, 2005; Kremer et al., 2005; Alsteens et al., 2008). Formally, the cell envelope is composed of an outer capsule-like layer, a complex cell wall, and the plasma membrane, which

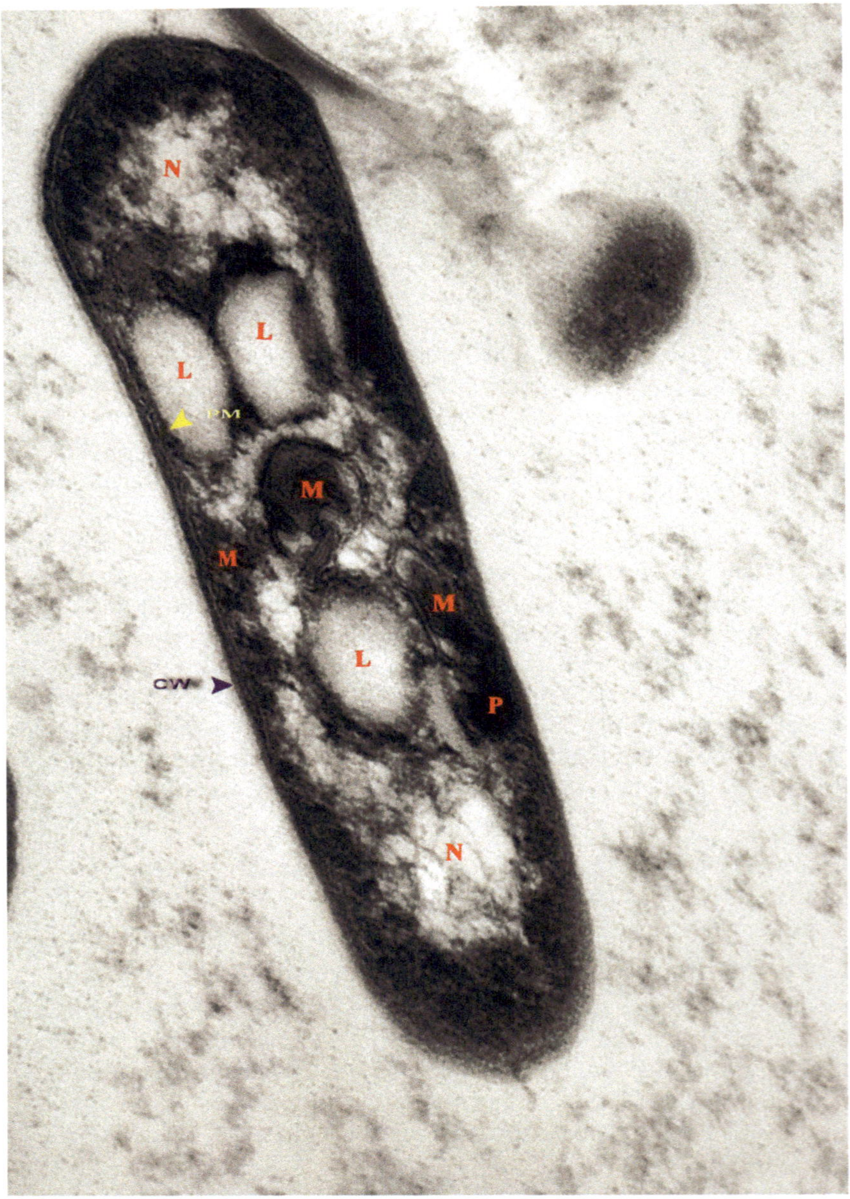

FIGURE 2.1 Longitudinal thin section of *Mycobacterium tuberculosis*. The cell wall (CW), plasma membrane (PM), lipid vacuoles (L), mesosome (M), and nuclear material (N) are visible.

is apparently homologous to plasma membranes of other bacteria (Paul and Beveridge, 1994; Daffé and Draper, 1998; Hoffmann et al., 2008). The outer capsule-like structure contains proteins, polysaccharides, and minor amounts of inner lipids, which are apparently in constant turnover (Minnikin, 1982; Picard et al., 1984; Andersen et al., 1991; Romain et al., 1993; Laqueyrerie et al., 1995; Ortalo-Magné et al., 1995). This superficial layer, which occurs in all mycobacterial species, was described by Rastogi et al. (1986). Microscopic analysis of bacterial cell envelopes indicates that the layered architecture of the mycobacterial cell envelope is more related to that of gram-negative bacteria than that of classical gram-positive bacteria in which cytoplasm is surrounded by the plasma membrane and a thick peptidoglycan (PG) (Figs. 2.2–2.4).

The cell wall consists of a covalently linked cell wall skeleton (CWS) plus a somewhat heterogeneous collection of wall-associated molecules, lipids, and polypeptides (Kotani et al., 1959). The CWS (=13 nm) is chemically composed of three covalently linked constituents: PG (=9 nm), arabinogalactan (AG=4 nm), and mycolic acids (8 nm; Rastogi et al., 1986; Benedetti et al., 1984; Rastogi, 1991). Mycobacterial PG belongs to a family of structures possessed by almost all eubacteria (Draper, 1982). The structure is similar to one of the most common types found (eg, in *Escherichia coli*). However, it is slightly differs from the common type (Petit et al., 1969; Wietzerbin-Falszpan et al., 1970). One difference is the N-acylation of muramic acid with a glycol residue rather than the usual acetyl residue (Adam et al., 1969). The second difference concerns the occurrence of a substantial number of unusual cross-linkages between two chains of PG. In addition to the usual D-alanyl-diaminopimelate linkages, a proportion of bonds involving two residues of diaminopimelic acid have been characterized (Wietzerbin et al., 1974).

AG is composed of three arabinan chains (composed of ~27 D-arabinofuranosyl units each) attached to the homogalactan core consisting of 32 D-galactofuranosyl units of linear alternating 5- and 6-linked β-D-galactofuranosyl residues. The homoarabinan chains are composed of linear α-D-arabinofuranosyl residues with branching produced by 3,5-linked α-D-arabinofuranosyl units substituted at both positions by α-D-arabinofuranosyl residues. The nonreducing termini of the arabinan chains consist of pentaarabinosyl units; these units contain all of the 2-linked arabinofuranosyl residues and are the principal seroactive epitopes of the entire AG (Misaki et al., 1974; Daffé and Draper, 1998).

In the CWS, PG is in covalent attachment to AG, which in turn is attached to the mycolic acids. This is termed the cell wall core of the mycolylarabinogalactan PG complex (Liu and Gotschlich, 1967; Cunto et al., 1969). Mycolic acids are long-chain, α-branched, β-hydroxy fatty acids (Asselineau and Lederer, 1950; Minnikin and Polgar, 1966). Mycobacterial strains generally contain a complex mixture of different types of mycolates that are readily fractionable by chromatography (Daffé et al., 1983; Minnikin et al., 1984; Etémadi et al., 1966). Mycobacterial mycolates may be grouped into structural types according to the absence or presence of defined oxygen functions in addition to the common hydroxyl (on C3) and carboxyl (C1) groups;

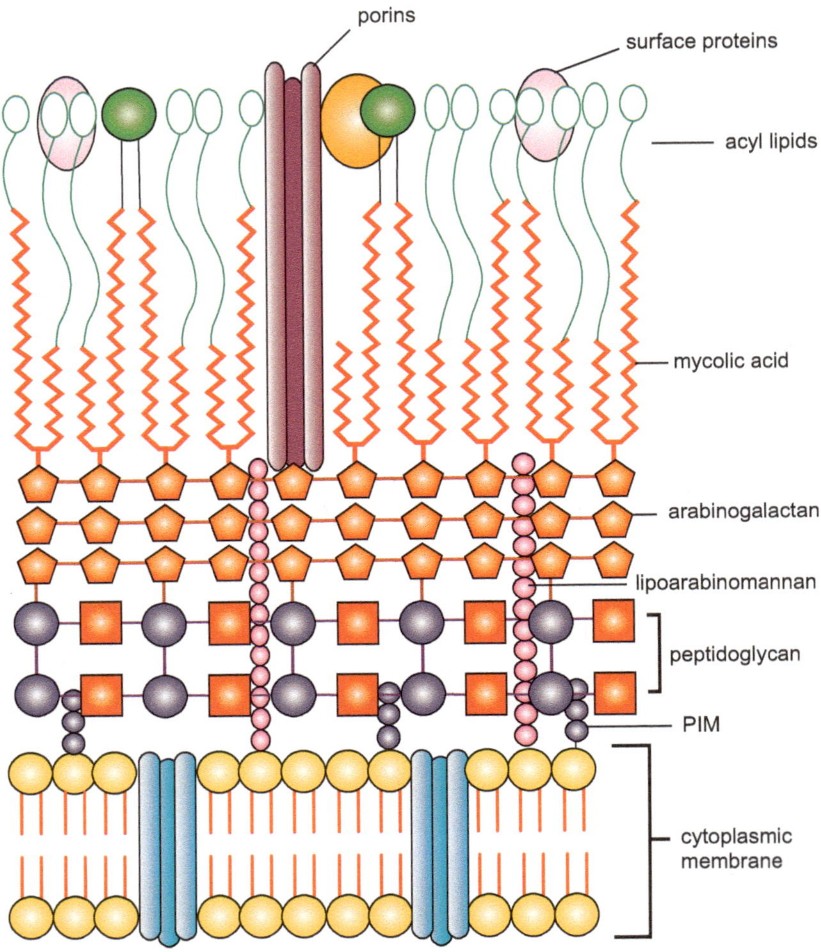

FIGURE 2.2 Schematic diagram of *Mycobacterium* cell wall. The proposed arrangement is cytoplasmic membrane, periplasm, cell wall compounds (peptidoglycan, arabinogalactan, mycolic acids) and associated lipids, and outer layer (called capsule). Similar to the outer membrane of the gram-negative cell wall, porins are required to transport small hydrophilic molecules through the outer membrane of the acid-fast cell wall. *PIM*, Phosphatidylinositol mannosides.

these functions include keto, methoxy, epoxy, and carboxy groups. The mycolic acids in *Mycobacterium tuberculosis* are alpha, keto, and methoxymycolates containing 76–82, 84–89, and 83–90 carbons, respectively (Etémadi,1967; Minnikin, 1982) (Figs. 2.5 and 2.6).

The upper segment of the cell wall (outer layer = 12 nm) is composed of free lipids—some with longer fatty acids complementing the shorter α-chains and some with shorter fatty acids complementing the longer chains. Interspersed somehow are the cell wall proteins, the phosphatidylinositol mannosides (PIMs),

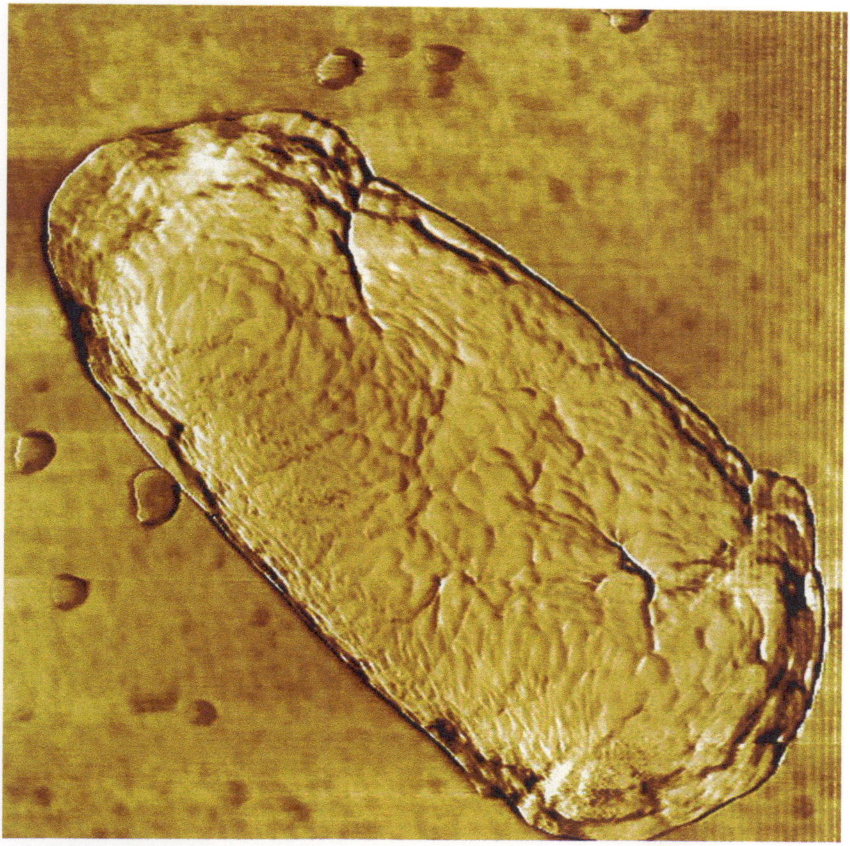

FIGURE 2.3 *Mycobacterium tuberculosis* imaged by atomic force microscopy. The cell surface is homogenous and rough: the root mean square roughness is ~8.0 ± 2.5 nm. The roughness represents an electron-dense outer layer and not the electron-dense zone.

the phthiocerol-containing lipids, lipomannan (LM), and lipoarabinomannan (LAM). Traversing the whole envelope, some glycolipids, such as the phosphatidylmyoinositol mannosides, LM, and LAM are anchored to the plasma membrane and extended to the exterior of the cell wall (Figs. 2.7 and 2.8).

The plasma membrane or cytoplasmic membrane (= 7 nm) has a triple-layered profile characteristic of biomembranes (Brodie et al., 1979). The membrane has an asymmetric profile with the outer layer (facing the cell wall) thicker and denser than the inner layer (facing the cytoplasm). These layers are an electron-dense layer that is separated by an electron-transparent zone. The outer layer is composed of mannose-containing phospholipids (PIMs) of approximately 3 nm and reacts positively by the cytochemical method for saccharides of Thiéry (Rastogi, 1991); the inner layer is approximately 2 nm. The membrane does not

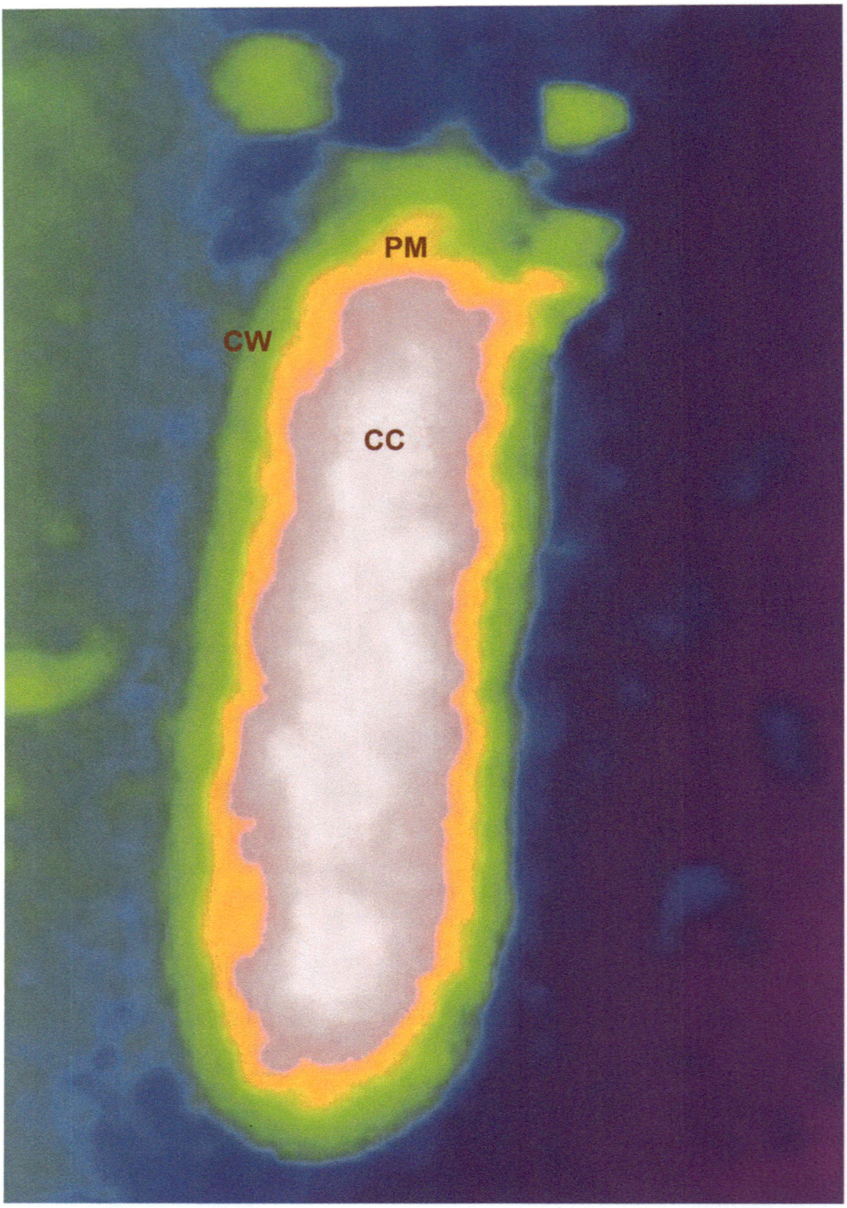

FIGURE 2.4 Image recorded by the height mode of atomic force microscopy. The three predominate parts of the *Mycobacterium tuberculosis* envelope are the cell wall (CW), plasma membrane (PM), and cytoplasmic content (CC).

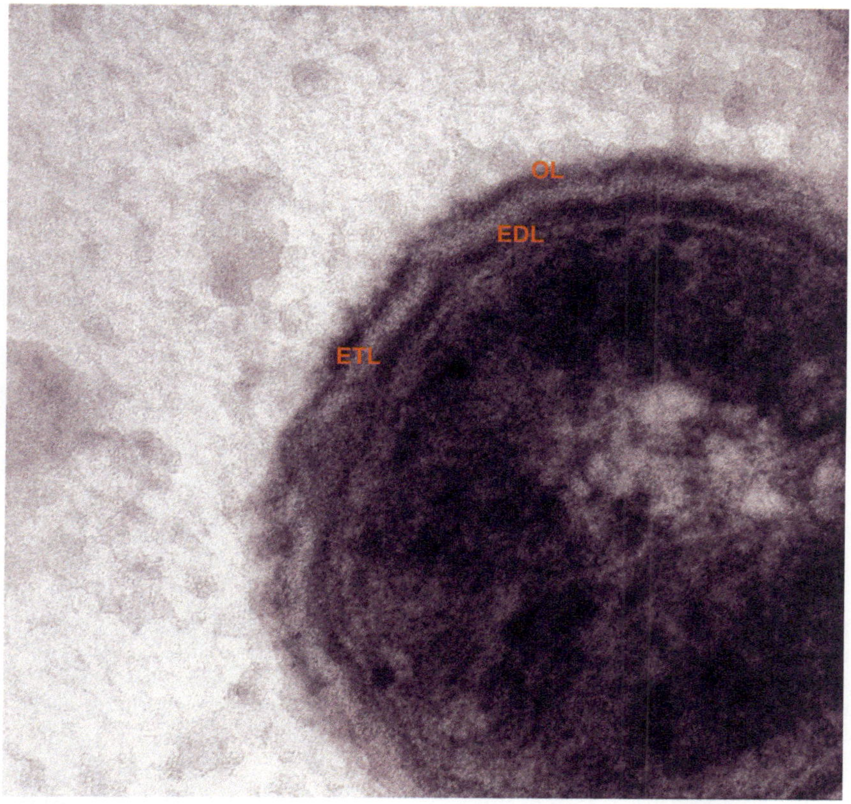

FIGURE 2.5 Thin cross-section of cell wall of *Mycobacterium tuberculosis*. From cytoplasm to the cell periphery, a unit plasma membrane bilayer is followed by an electron-transparent periplasmic space, an electron-dense peptidoglycan layer an electron-transparent middle wall an electron-dense outer layer (OL), and an electron-transparent zone that surrounds the OL.

have any peculiar characteristics except for the presence of some lipopolysaccharides that are shared by all Actinomycetales (Mahapatra et al., 2005). The membrane provides osmotic protection, regulates the traffic of specific solutes between the cytoplasm and the environment, and assumes the cell housekeeping tasks. The cytoplasmic membrane is surrounded by a cell wall that protects the cell contents, provides mechanical supports, and is responsible for the shape-forming property and the structural integrity of the bacterium (Akamatsu et al., 1966; Oka et al., 1968; Goldman, 1970; Kumar et al., 1979) (Figs. 2.9–2.16).

The cytoplasmic content of *M. tuberculosis* is composed of ribosomes, nucleoids, and inclusions. Ribosomes are present in the typical dimensions of the prokaryotic ribosome. Their number per cell varies; they seems to disappear totally from old cells (4 weeks or older; Koike and Takeya, 1961; Trnka et al., 1968; Worcel et al., 1968; Hirata, 1979).

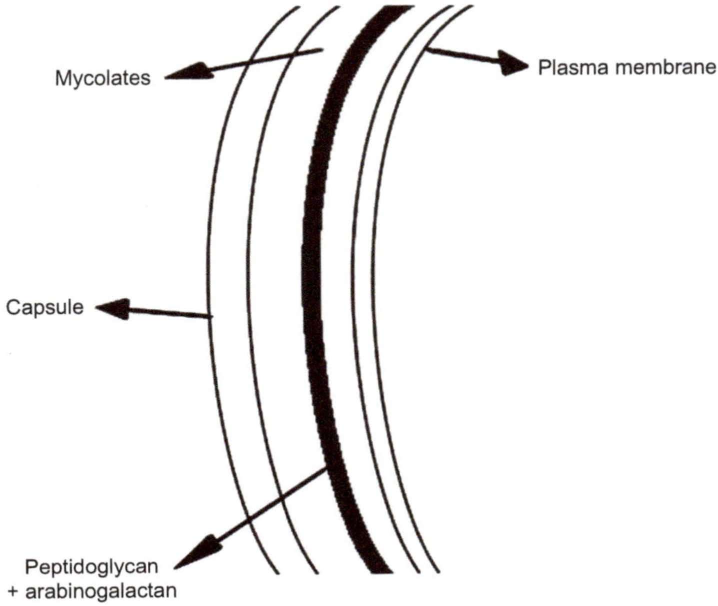

FIGURE 2.6 The arrangement of the various layers is diagrammatically shown. The innermost layer is the plasma membrane. Outside of the plasma membrane is the cell wall skeleton, which consists of peptidoglycan and arabinogalactan followed by mycolate residue. The surface of the envelope consists of a more or less loosely attached capsule-like layer of polysaccharide and protein, with rather little lipid (Draper, 1998).

The nucleoid is made up of nonaggregated DNA fibrils, messenger RNA, and transcription factor proteins. Inclusions or nonliving substances are present in *Mycobacterium* cytoplasm and include polysaccharides and lipids. They are stored nutrients and secretory products (Silva and Macedo, 1982) (Figs. 2.17–2.23).

The permeability of mycobacteria to substances in their environment is controlled by the properties of their envelopes (Daffé and Reyrat, 2008). The unique mycolic acid bilayer acts as an efficient barrier that can be the major determinant of the intrinsic resistance of mycobacteria to most common antibiotics, chemotherapeutic agents, and chemical disinfectants (Brennan and Nikaido, 1995). The development of unusual structure(s) in the mycobacterial cell envelope has been highlighted. This specific pathway can transport the proteins and polysaccharides across the outer lipid barrier (Kartmann et al., 1999; Alteri et al., 2007; Velayati et al., 2009a,b; Danilchanka et al., 2014; Hosseini et al., 2014).

One such pathway is formation of water-filled protein channels, the porins (Nikaido, 1994a,b). Porins are transport proteins that allow the direct diffusion of small and hydrophilic molecules into the periplasm (Trias et al., 1992; Trias and Benz, 1993). This sort of pore was already well recognized and studied in gram-negative organisms, where it is formed by specialized outer-membrane

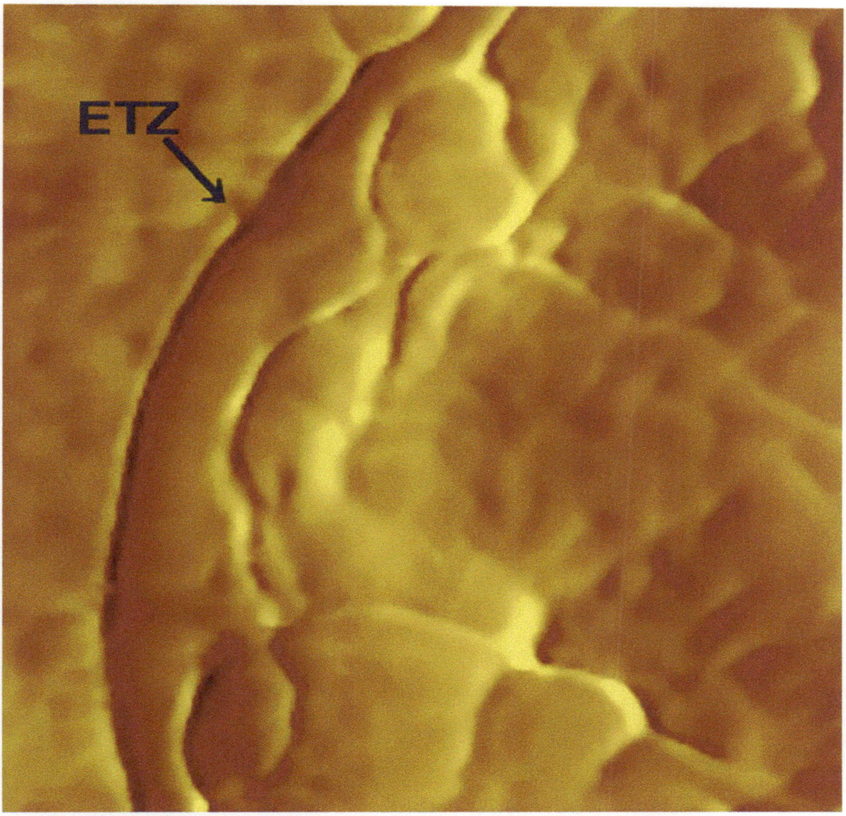

FIGURE 2.7 Vertical cross-section of *Mycobacterium tuberculosis* cell wall. Atomic force microscopy observation is consistent with transmission electron microscopy data, which revealed a multi-layered organization of the cell wall envelope. The electron-transparent zone (ETZ) is clearly visible.

proteins (Nikaido, 1994a,b). A search was performed for analogous pore-forming proteins in *M. tuberculosis*. Kartmann et al. (1999) described two porins in *M. tuberculosis.* One of them is composed of 15-kDa subunits and has a channel conductance of 0.7 nS. The other porin is a 60-kDa protein with a conductance of 3 nS. On the basis of nucleic acid sequence homology, Senaratne et al. (1998) identified the pore-forming outer membrane protein (OmpATb) from *M. tuberculosis.* OmpATb has a pore diameter of 1.4–1.8 nm and a single-channel conductance of only 0.7 nS. In general, the porins of *M. tuberculosis* are of particular interest because three of four first-line tuberculosis drugs are hydrophilic and understanding the porin pathway would promote the design of new drugs to fight tuberculosis (Raynaud et al., 2002; Niederweis, 2003; Veyron-Churlet et al., 2011; Yang et al., 2011) (Figs. 2.24–2.46).

The observation of rope-like filaments on the surface of mycobacteria by electron microscopy was first reported by Draper and Rees (1970). Their studies reported the

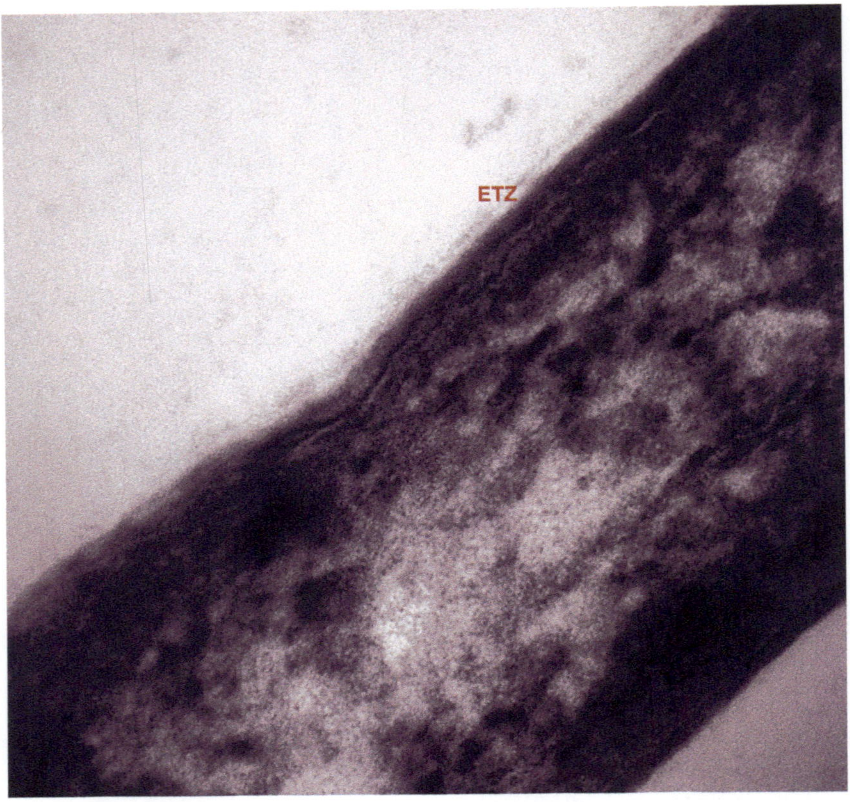

FIGURE 2.8 Electron-transparent zone (ETZ) in longitudinal section of *Mycobacterium tuberculosis* cell wall. ETZ represents an unbound pseudocapsule. The capsule consists of polysaccharides, proteins, and glycolipids. The capsule may be protective and bioactive.

appearance of the fibers as 7 nm in diameter, which seemed similar in morphology to isolated waxes or mycolic acid derived from the cell wall of mycobacteria. Later, Barksdale and Kim (1977) published a series of papers describing the superficial mycosidic integuments of mycobacteria. Barrow et al. (1980) described the superficial fibrillar material as 4 nm in diameter with a curious transparent appearance. The isolated fibers were determined to be composed of lipids, predominantly polar peptidoglycolipids. In most cases the observed fibers were attributed to C-mycosides or peptidoglycolipids derived from the cell envelope. These fibers or pili can mediate the specific interaction of mycobacteria with a target in the host (Strom and Lory, 1993; Klemm and Schembri, 2000). The occurrence of pili was shown in laboratory strains ($H_{37}R_V$ and CDC1551; Alteri et al., 2007) and clinical isolates of *M. tuberculosis* (Velayati et al., 2009a,b, 2012) (Figs. 2.47–2.52).

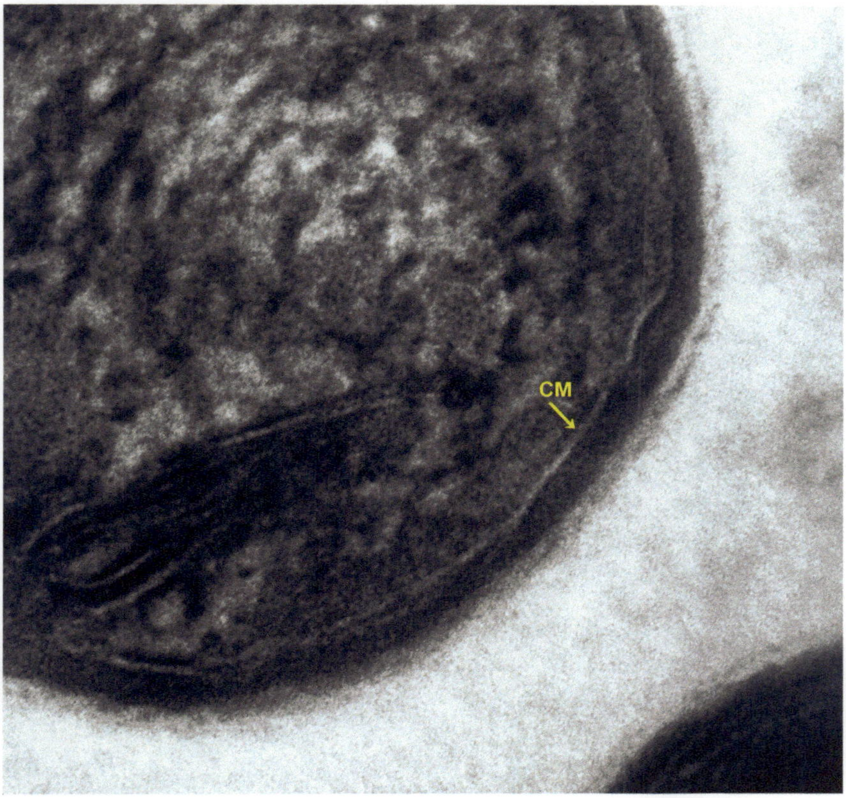

FIGURE 2.9 The cytoplasmic membrane (CM) has an asymmetric profile with the outer layer (facing the cell wall) being thicker and denser than the inner layer (facing the cytoplasm).

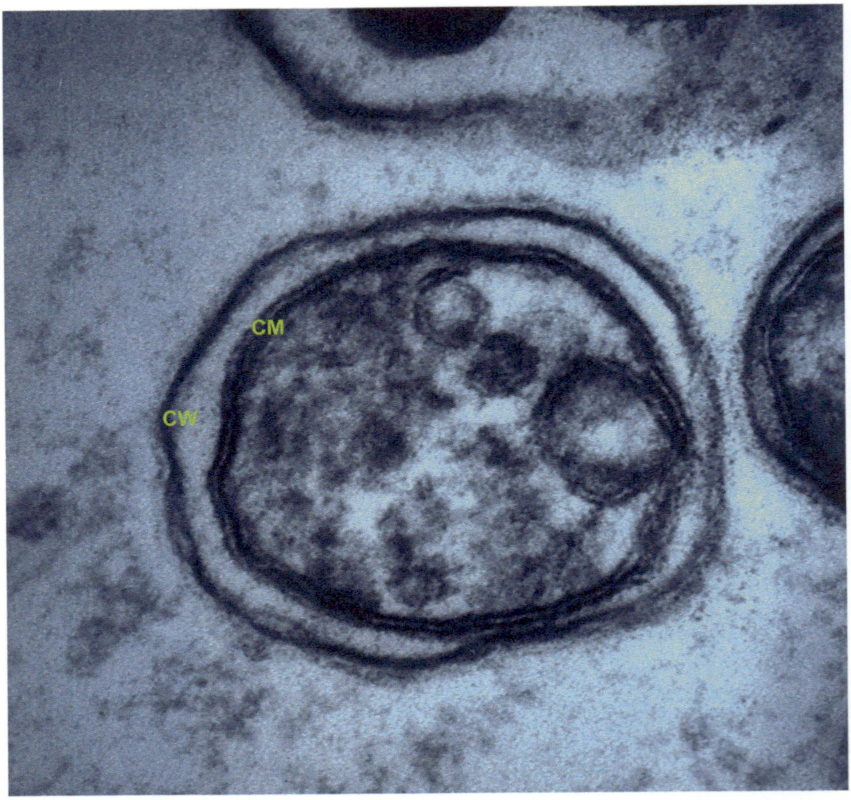

FIGURE 2.10 Cross-section of heat-killed *Mycobacterium tuberculosis*. The cell wall (CW) and a cytoplasmic membrane (CM) are continuous without deformations or reduction in thickness.

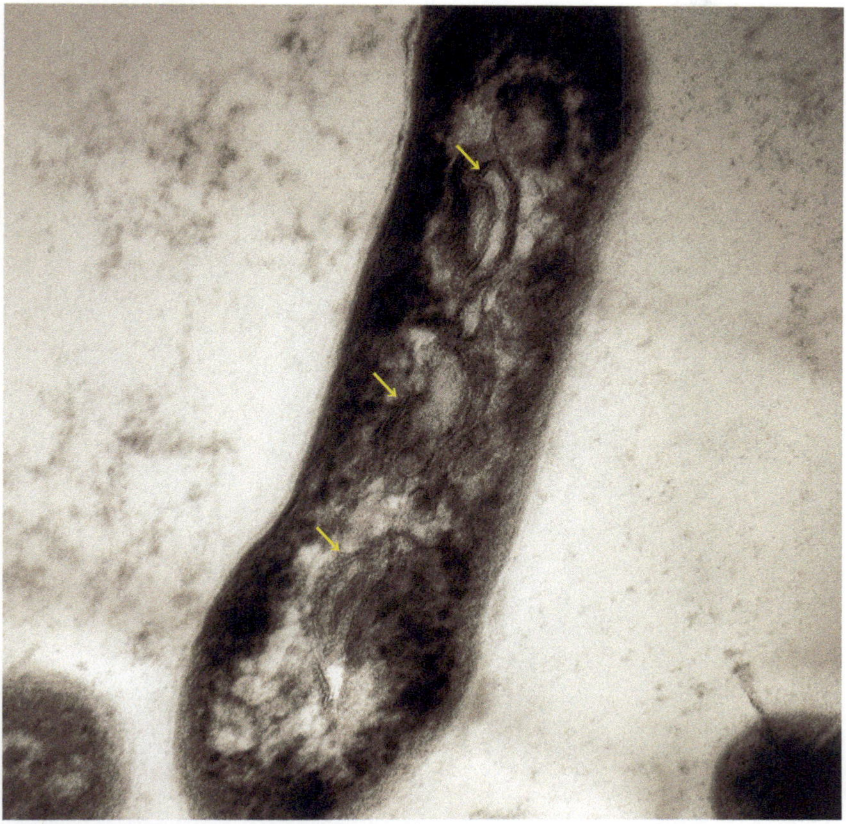

FIGURE 2.11 The cytoplasm contains a complicated intracytoplasmic membrane system, so-called mesosomes (*yellow arrows*). Mesosomes appear as a result of invagination and subsequent bifurcation of the cytoplasmic membrane. On longitudinal sections, mesosomes can appear as vesicular, tubular, and/or lamellar elements within a membrane envelope, which is connected to the cytoplasmic micromembrane.

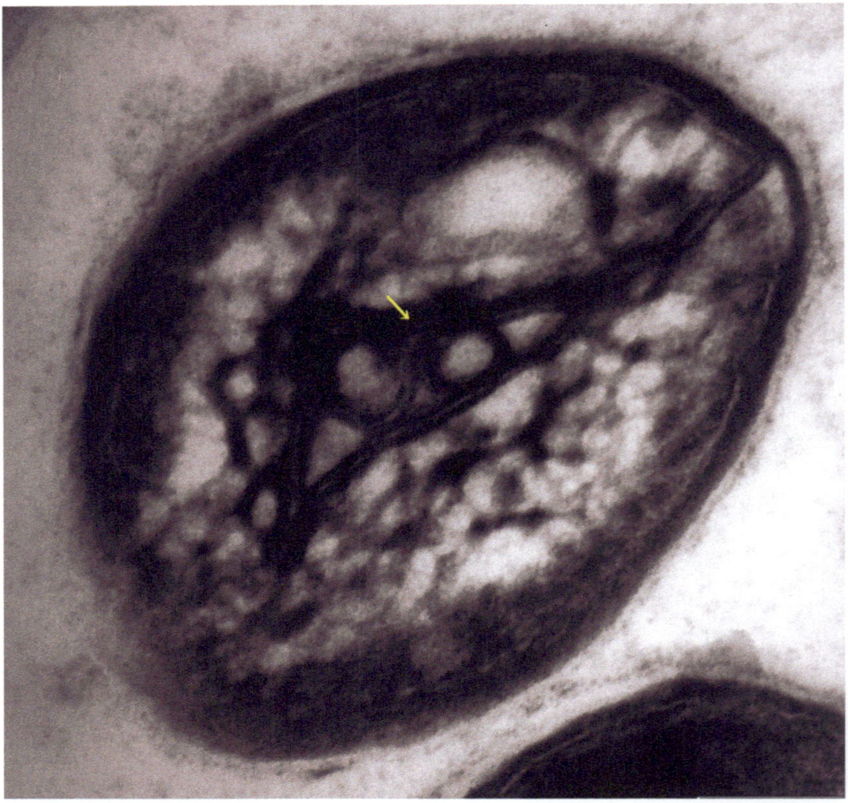

FIGURE 2.12 The expansion of the mesosome neck from the cytoplasmic membrane is demonstrated. The extension and concentration of the cytoplasmic membrane into the "lamellar structure" may serve in intensifying the metabolic activity of the bacteria (*yellow arrow*).

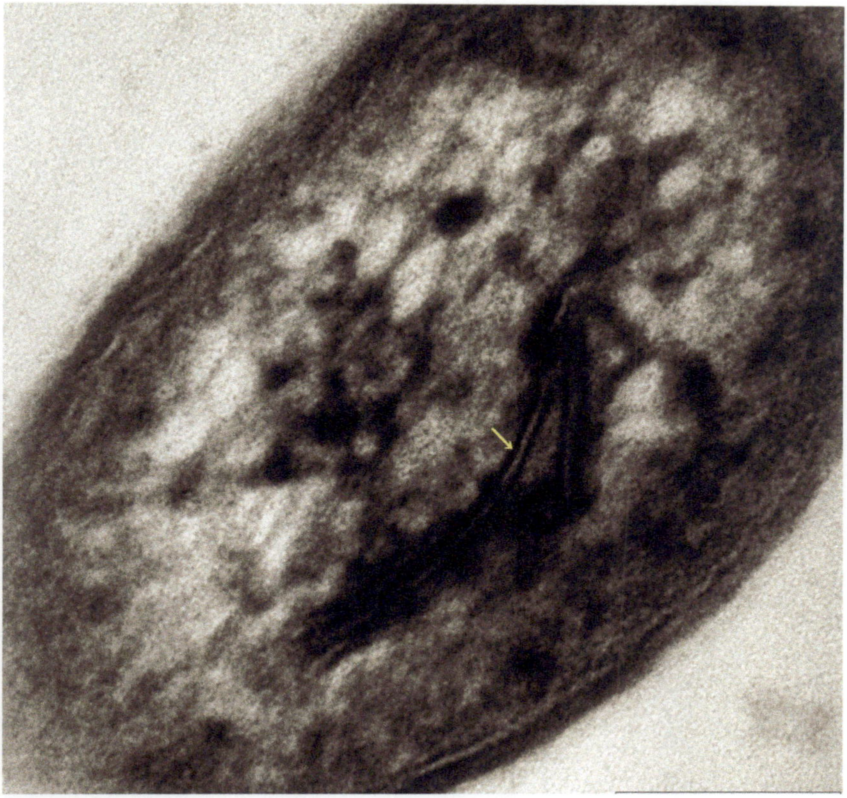

FIGURE 2.13 The mesosomal membrane systems are trilaminar, consisting of two electron-dense layers separated by an electron-transparent zone. The extension and concentration of the cytoplasmic membrane into the "lamellar structure" (*yellow arrow*) may suggest the phylogenetic origin of the membranous system.

FIGURE 2.14 The type and location of the mesosome within the cell may vary depending upon the physiological state of the cell. They are differentiated according to their cellular location: the septal mesosomes, peripheral or cytoplasmic membrane mesosome, and nuclear mesosomes (*arrows*).

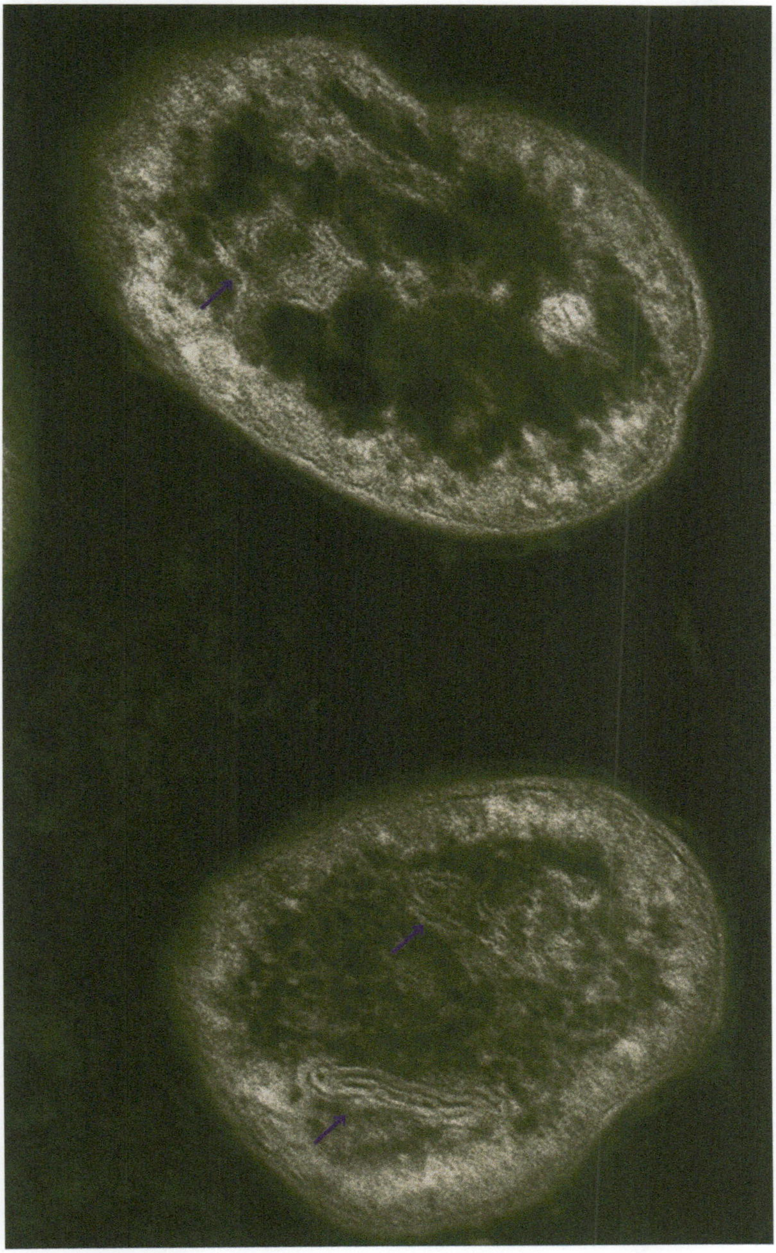

FIGURE 2.15 The tubovesicular, lamellar-vesicular, and lamellar-tubular mesosome are demonstrated in cross-sectional analysis (*arrows*).

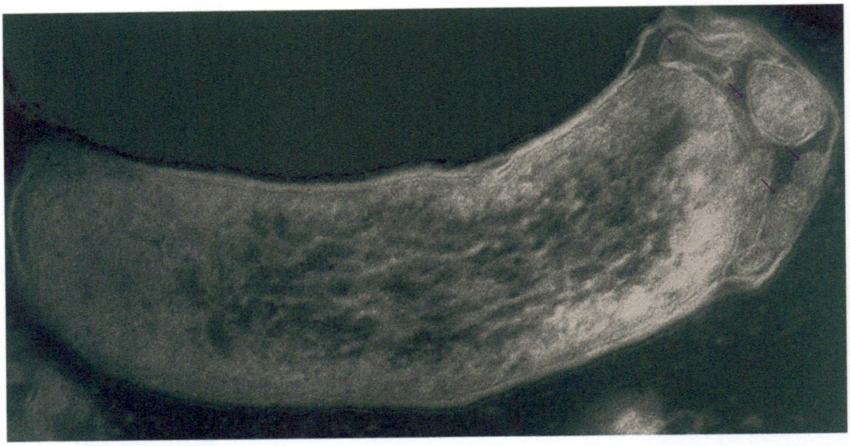

FIGURE 2.16 Accommodation of cytoplasmic membrane mesosome in the top of elongated *Mycobacterium tuberculosis* (*arrows*).

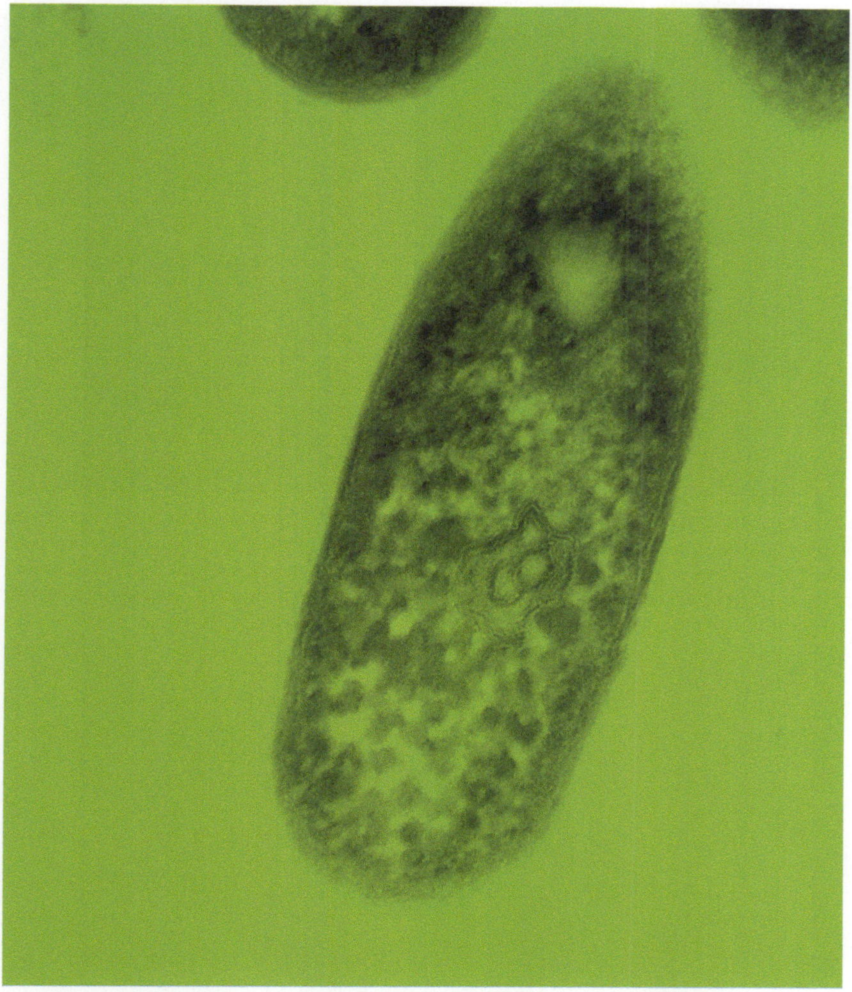

FIGURE 2.17 The cytoplasm or protoplasm is the portion of the cell that lies within the cytoplasmic membrane. The cytoplasm contains low-density areas with fine granular material or threads consistent with nuclear regions. Multiple small round to oval granules of various electron densities are the most prominent features of the cytoplasm.

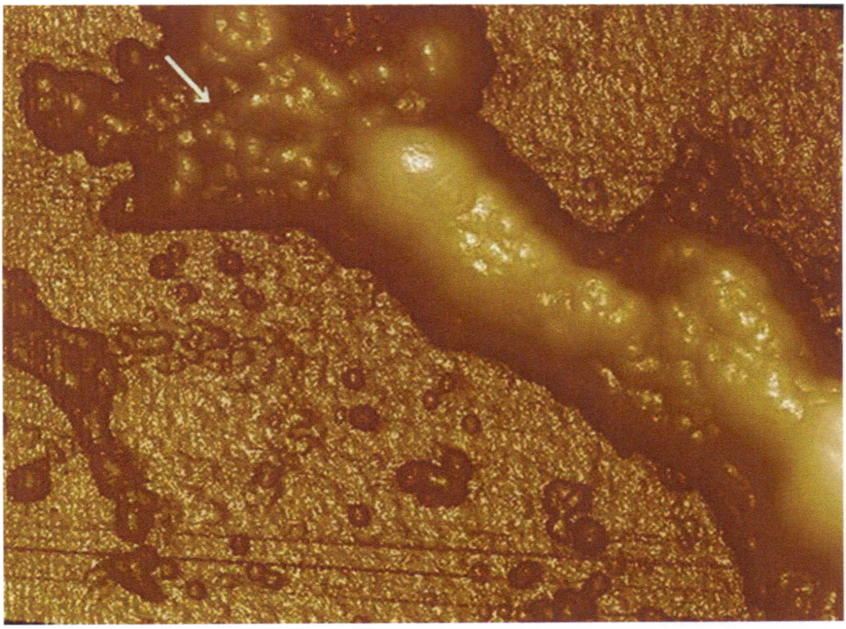

FIGURE 2.18 *Mycobacterium tuberculosis* discharges the cytoplasmic content. The cytoplasm itself is gel-like in consistency and includes chromosomes and ribosomes (*arrow*). The other constituents of cytoplasm include proteins, vitamins, ions, nucleic acids and their precursors, amino acids and their precursors, carbohydrates and their derivatives, and fatty acids and their derivatives.

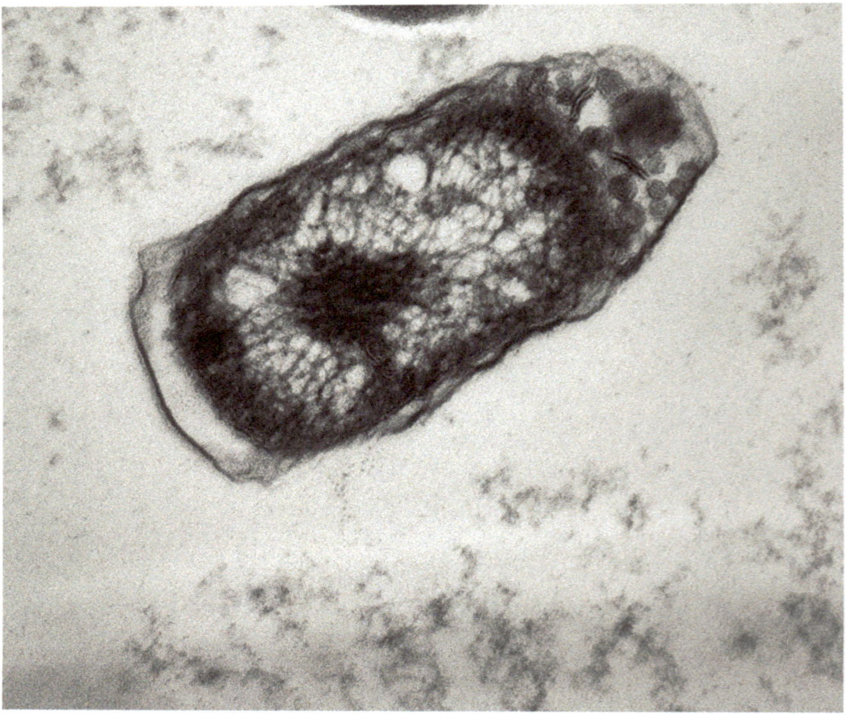

FIGURE 2.19 Cross-section shows nucleoid in *Mycobacterium tuberculosis*. The nucleoid appears as an electron-transparent zone containing numerous fine electron-dense fibrils that may form aggregations depending upon fixation.

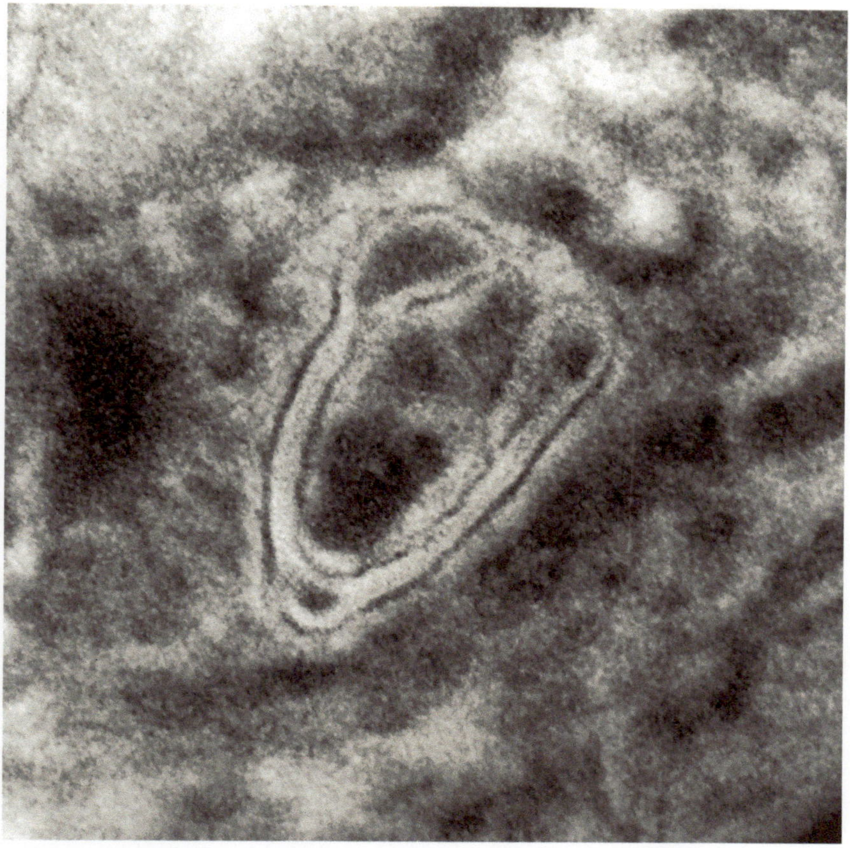

FIGURE 2.20 *Mycobacterium tuberculosis* has a single circular chromosome that is not enclosed in a specific membrane-surrounded compartment. The chromosome is attached to an invagination of the cytoplasmic membrane called the mesosome. It comprises 4,411,529 bp with approximately 3900 genes encoding proteins. The high guanine/cytosine content (G+C; 65.5%) was found to be uniform along most of the genome, confirming the hypothesis that horizontal gene transfer events are virtually absent in modern *Mycobacterium tuberculosis*.

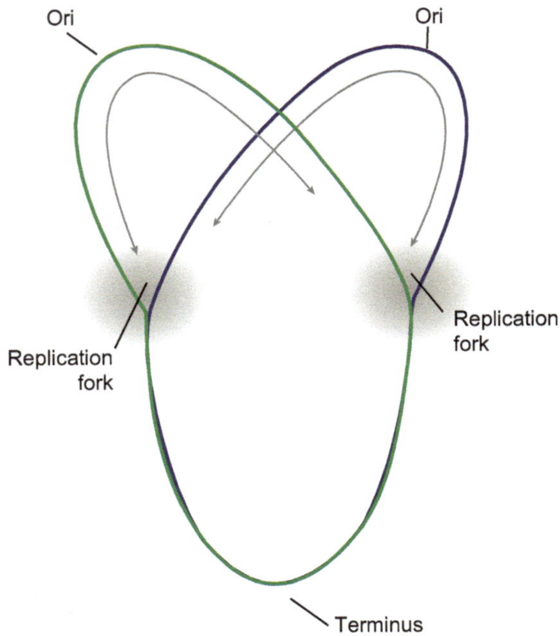

FIGURE 2.21 DNA replication in circular mycobacterial chromosome. The three-dimensional arrangement of the bacterial chromosome is shown, and the origin (Ori) and terminus of replication present at a very reproducible pattern of localization during the cell cycle. The origin region of the chromosome is generally AT nucleotide rich, containing repeats of AT-rich nucleotide sequences varying in length from 13 to 16, together with several 9-nucleotide DnaA protein recognition sequences called DnaA boxes.

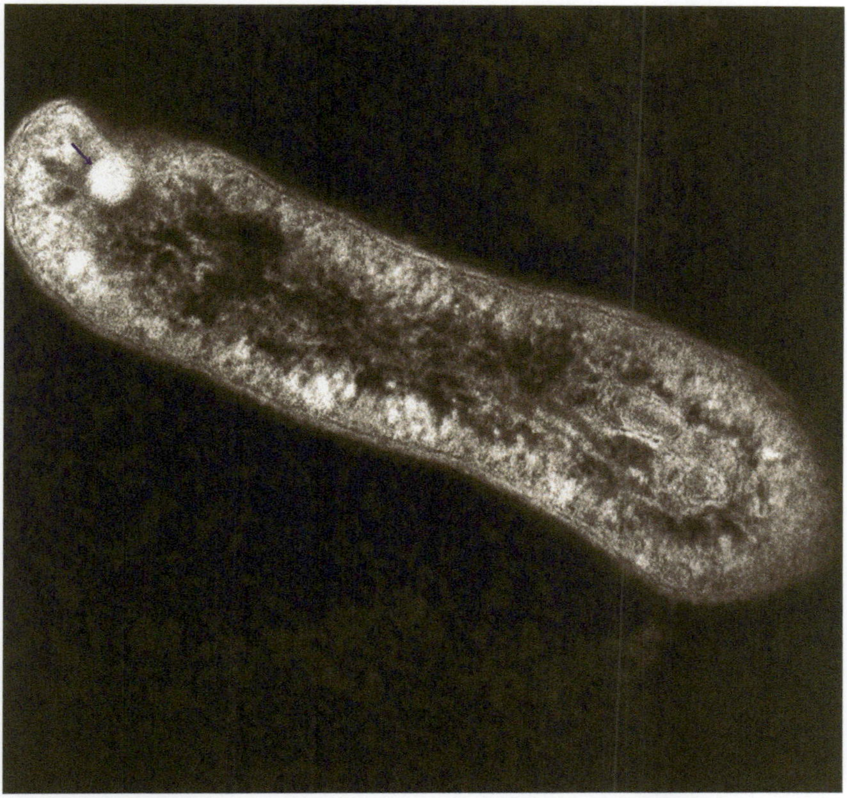

FIGURE 2.22 Polyphosphate granules (*arrow*) are present inside of the cytoplasm and are not membrane associated. Polyphosphate granules are associated with diverse functions including energy storage and mycobacterial stringent response.

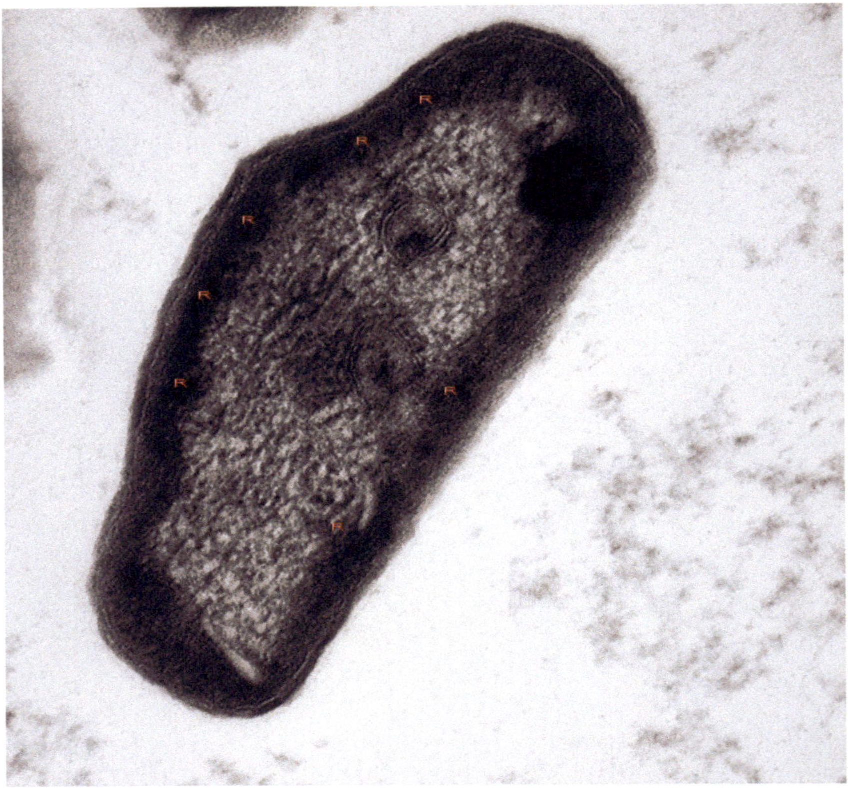

FIGURE 2.23 Ribosome (R) as electron-dense particles (10–20 nm in diameter) in the cytoplasm of the cell. The average number of ribosome in each cell is 1.672 ± 568, and the ribosome density was 716.5 ± 171.4/0 ft.

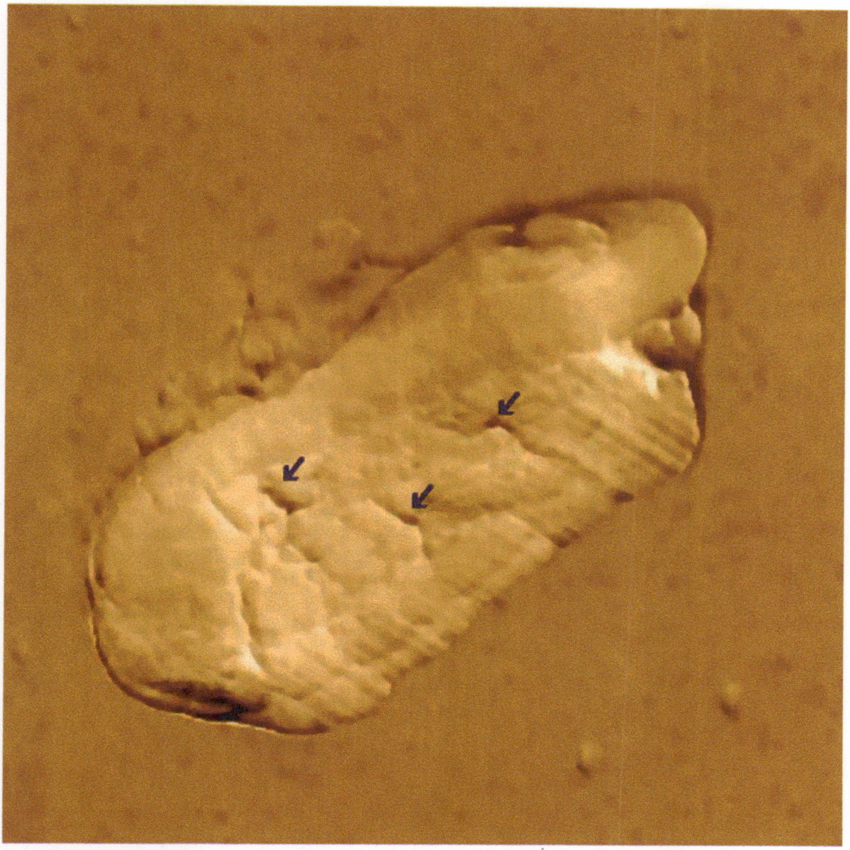

FIGURE 2.24 Whole cell of *Mycobacterium tuberculosis* with porins on their surface. The pathogenicity of tubercle bacilli is largely dependent on its surface structures. Among the components of the outer membrane, the outer membrane proteins, such as porins (*arrows*), can play a fundamental role in pathogenicity and represent useful targets for therapeutic development.

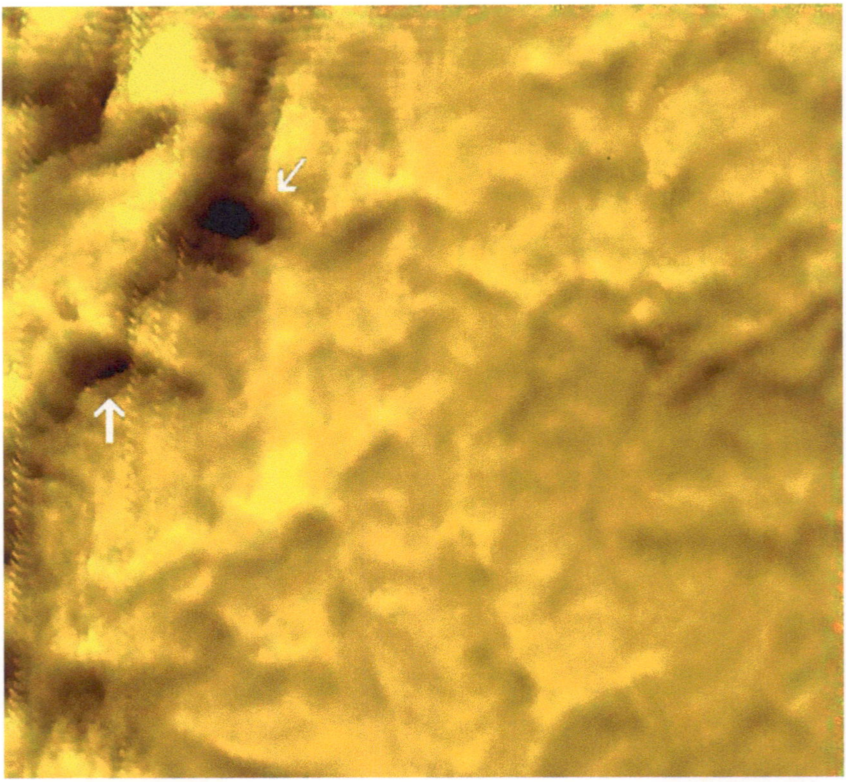

FIGURE 2.25 Vertical section of *Mycobacterium tuberculosis* cell wall shows the porins (*arrows*). Porins or outer membrane proteins can be formed as a channel or pores in *M. tuberculosis*. The primary function of porins is to convert the outer membrane into a molecular sieve. Thus solutes too large to fit through the pores are excluded from passage. Recently, porins are proposed to be useful for uptake of antibiotics.

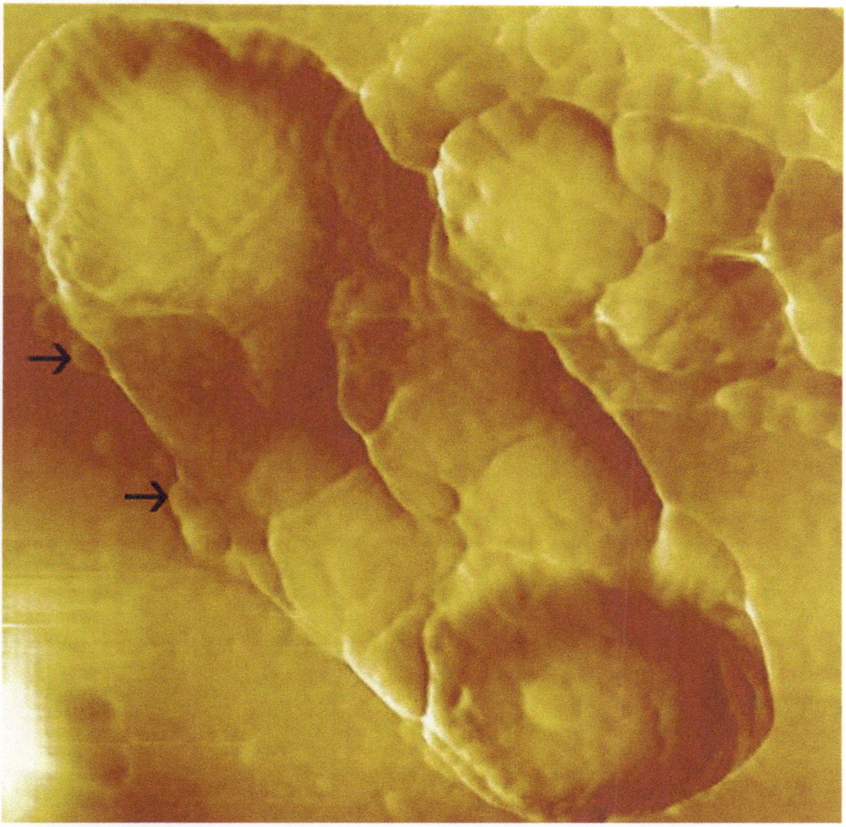

FIGURE 2.26 Whole cell of *Mycobacterium tuberculosis* recorded by atomic force microscopy. The channel-forming pores are seen on the side of *Mycobacterium tuberculosis (black arrows)*. Usually *Mycobacterium tuberculosis* adapts to low pH, low nutrients, and low oxygen. These channel-forming pores may be used for uptake of nutrients or discharge of toxic compounds, which requires further investigation.

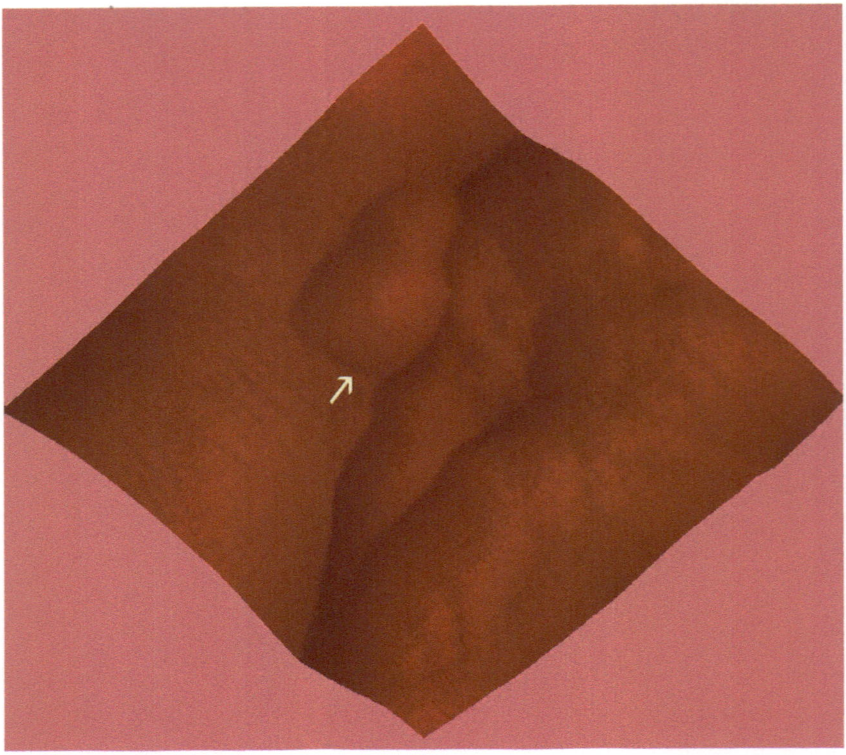

FIGURE 2.27 Vertical section of *Mycobacterium tuberculosis* shows the channel-forming pores recorded by atomic force microscopy (*arrow*).

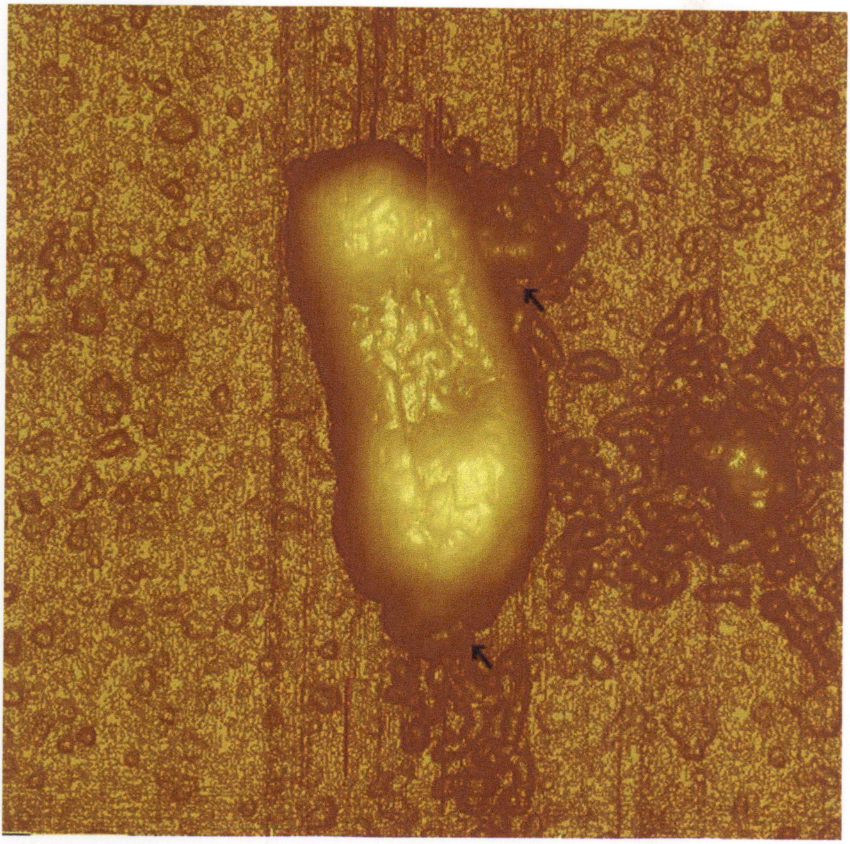

FIGURE 2.28 Cytoplasmic or protein discharge from channel-forming porins in *Mycobacterium tuberculosis (black arrows)*. Gram-negative bacteria use complex nanomachines such as Type I–VI secretion systems to secret effector proteins mediating host cell death and subverting the immune system. Likewise, there should be a secretion system capable of translocating the proteins across the membranes in *M. tuberculosis*.

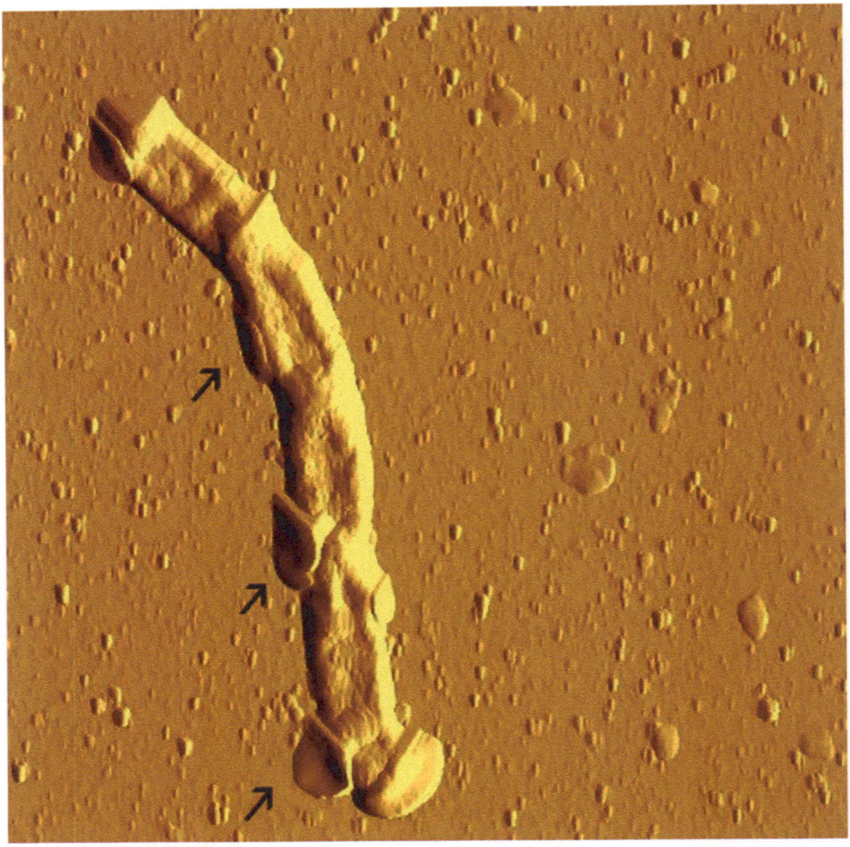

FIGURE 2.29 Outer membrane channel in totally drug-resistant *Mycobacterium tuberculosis*. Of these bacilli, 10–15% develop the outer membrane channels that are visible on their cell surface (*black arrows*). The function of such channels remains to be resolved.

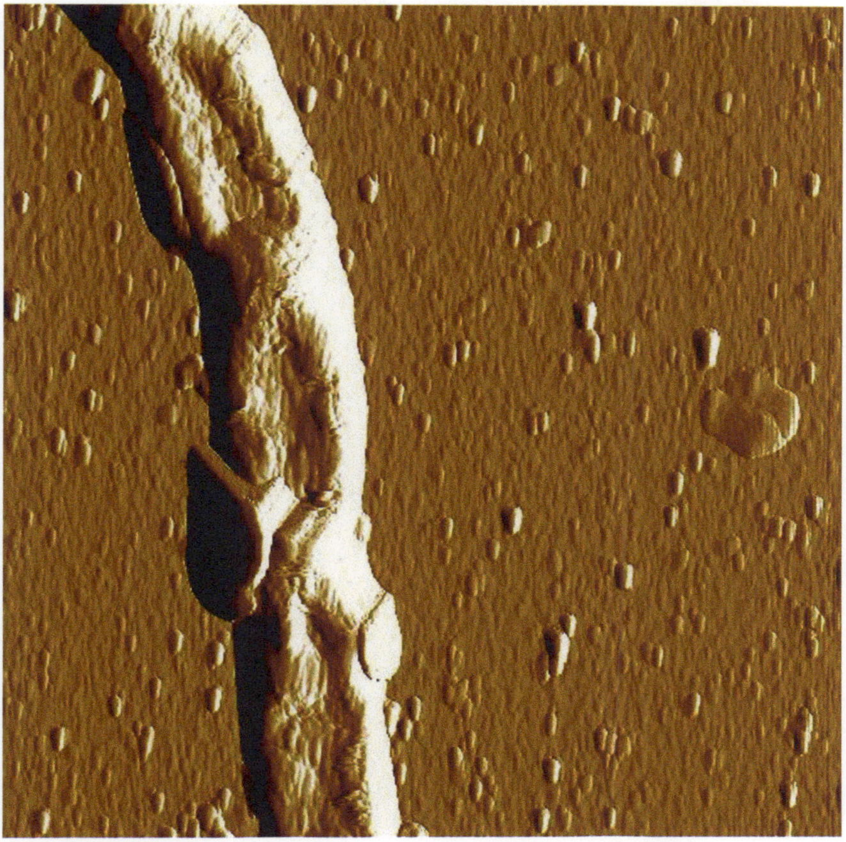

FIGURE 2.30 Outer membrane channel in side wall of totally drug-resistant *Mycobacterium tuberculosis* recorded by atomic force microscopy. The average diameter is from 1.4 to 3 nm.

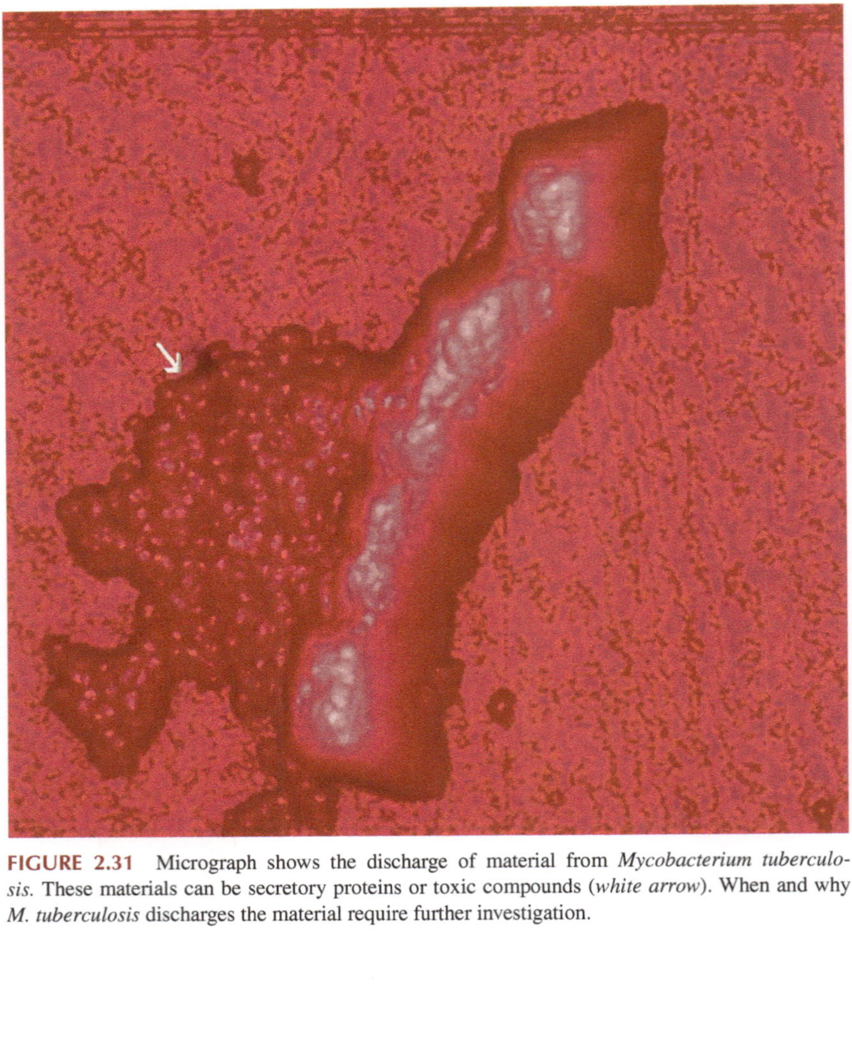

FIGURE 2.31 Micrograph shows the discharge of material from *Mycobacterium tuberculosis*. These materials can be secretory proteins or toxic compounds (*white arrow*). When and why *M. tuberculosis* discharges the material require further investigation.

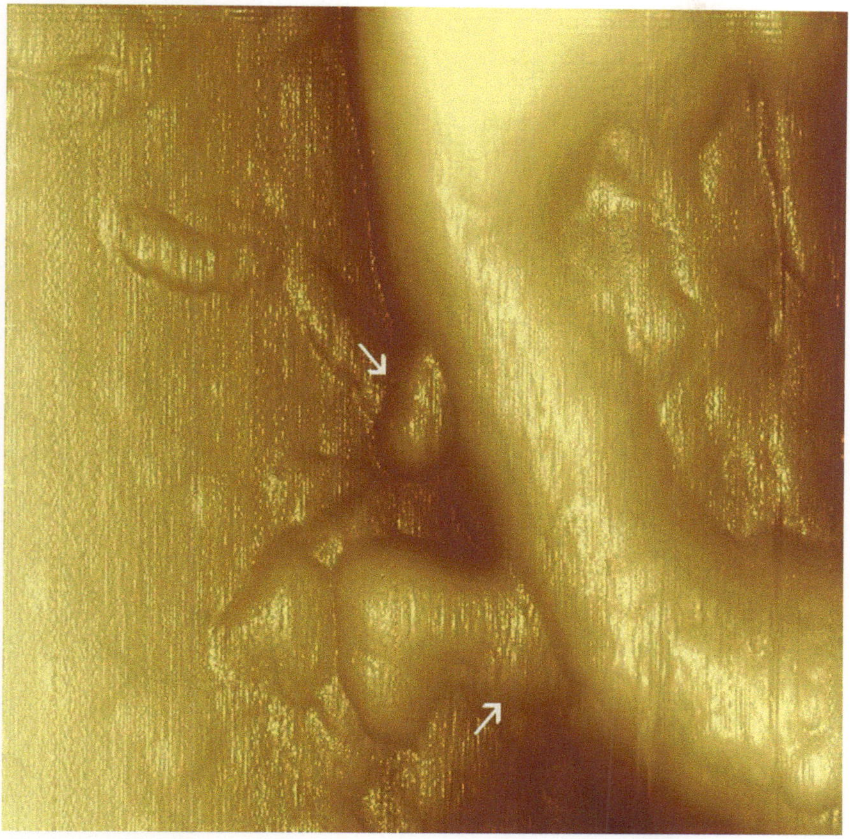

FIGURE 2.32 Side wall of *Mycobacterium tuberculosis* recorded by atomic force microscopy. The protrusion seems to be an outer membrane channel (*white arrows*).

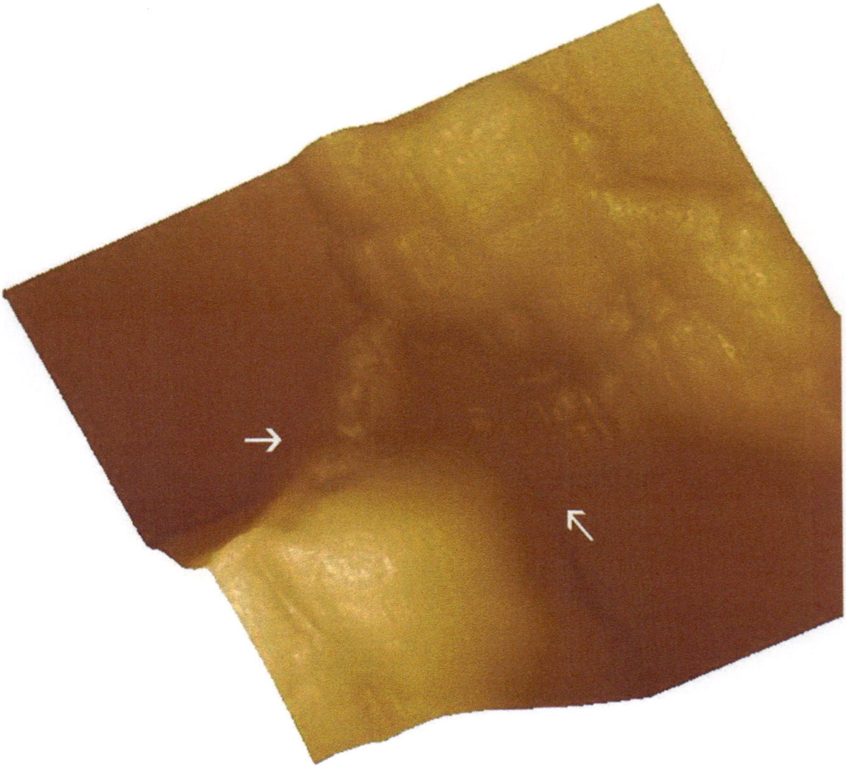

FIGURE 2.33 Side-wall protrusion in *Mycobacterium tuberculosis* cell wall at higher magnification scan. These channels can be used for discharge of materials (*white arrows*).

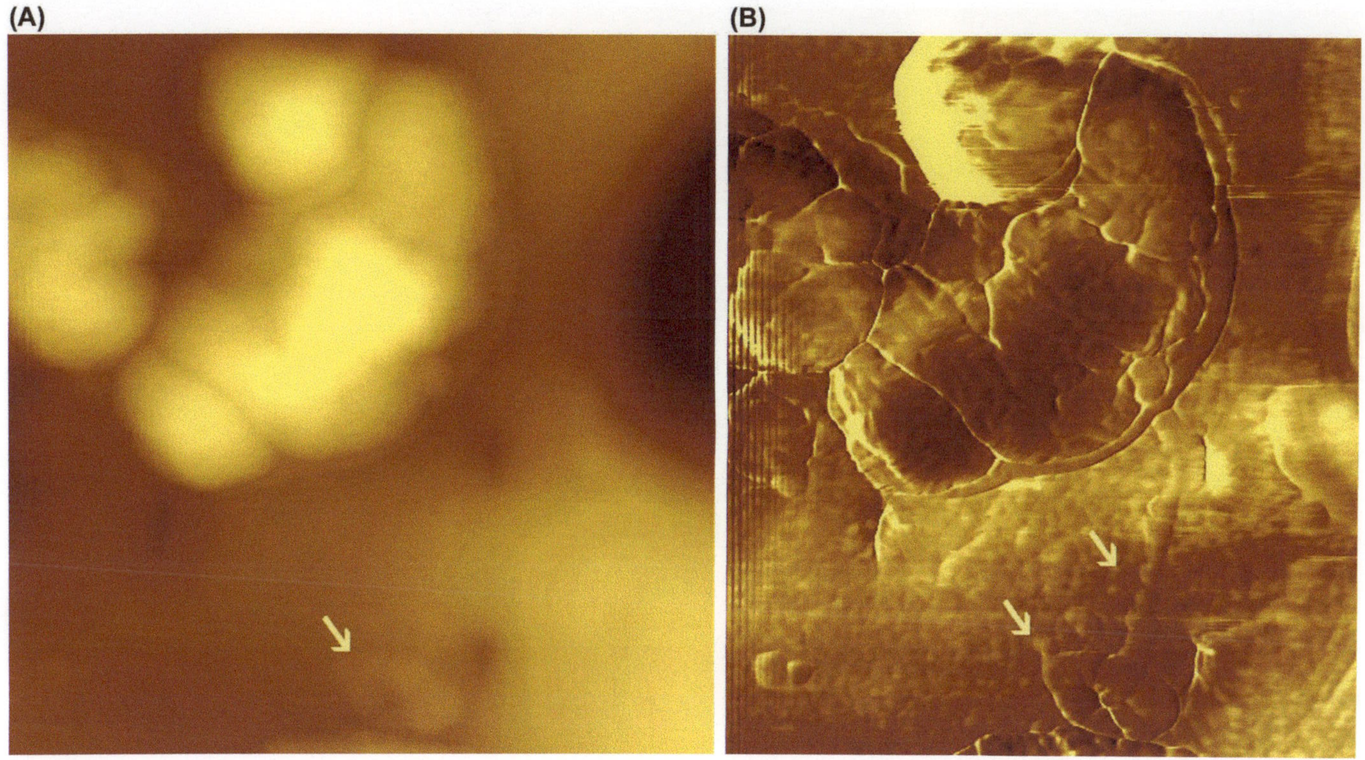

FIGURE 2.34 Tubular protrusion from totally drug-resistant tuberculosis under (A) height and (B) friction contact mode of AFM. The length of tubular extension reaches 890nm to 1.5μm. Their end is tapering (ie, a funnel-shaped tube) (*white arrows*). It is possible that *Mycobacterium tuberculosis* produces the outer membrane channel as a tubular structure on environmental constraints.

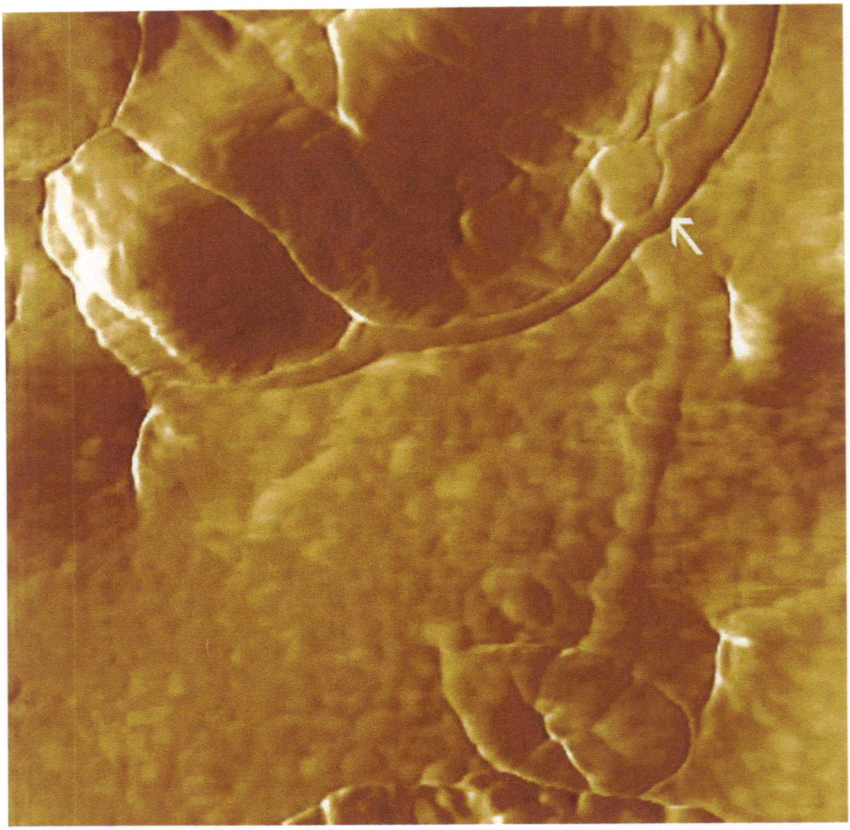

FIGURE 2.35 Cross-section of *Mycobacterium tuberculosis* with tubular protrusion (*white arrow*). The tubular protrusion originates from the cytoplasmic membrane. Two to four percent of totally drug-resistant tuberculosis strains elaborate such a long tubular extension.

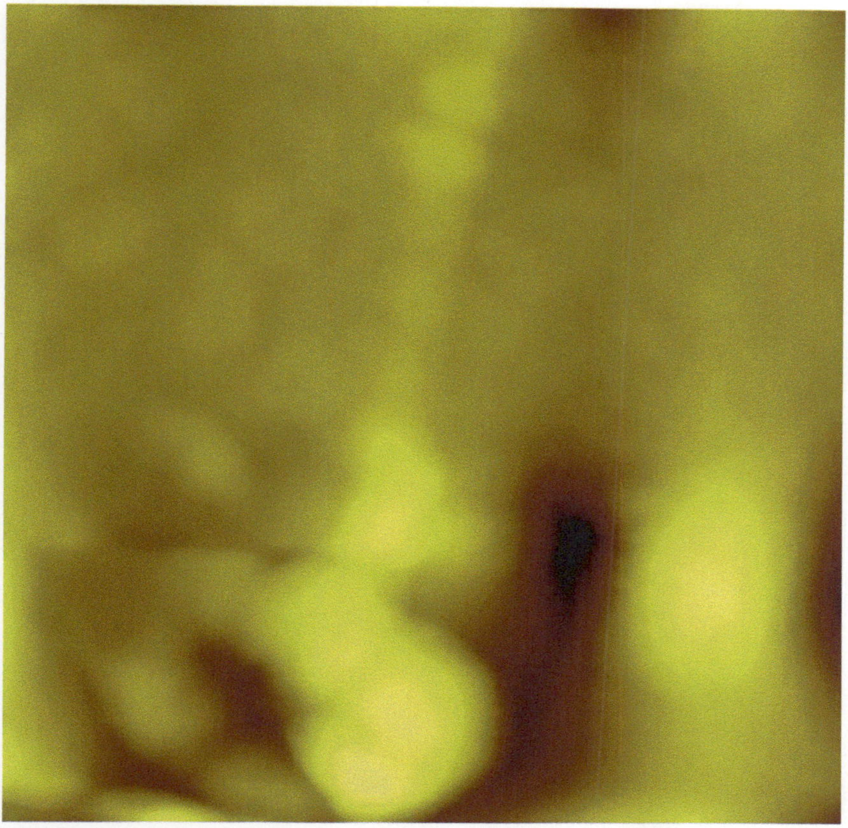

FIGURE 2.36 Height mode of atomic force microscopy shows the tubular protrusion. These structures had a low surface roughness (3.5 ± 1.2 nm), which highlight their origin from the cytoplasmic membrane.

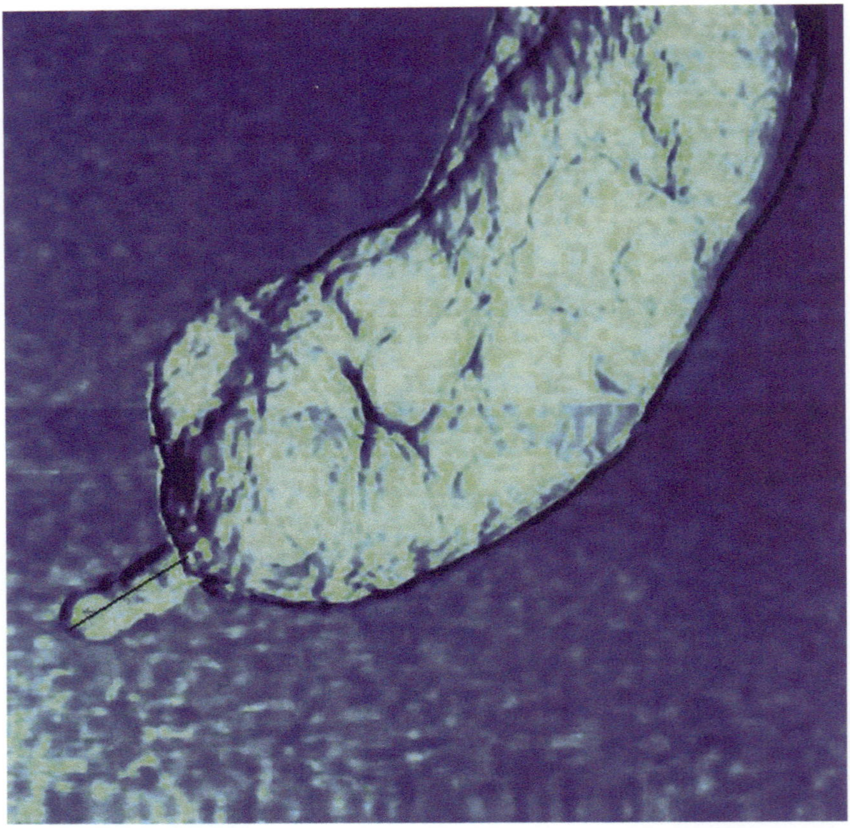

FIGURE 2.37 Tubular protrusion in susceptible *Mycobacterium tuberculosis*. The length of the tubular extension reaches to 20–70 nm.

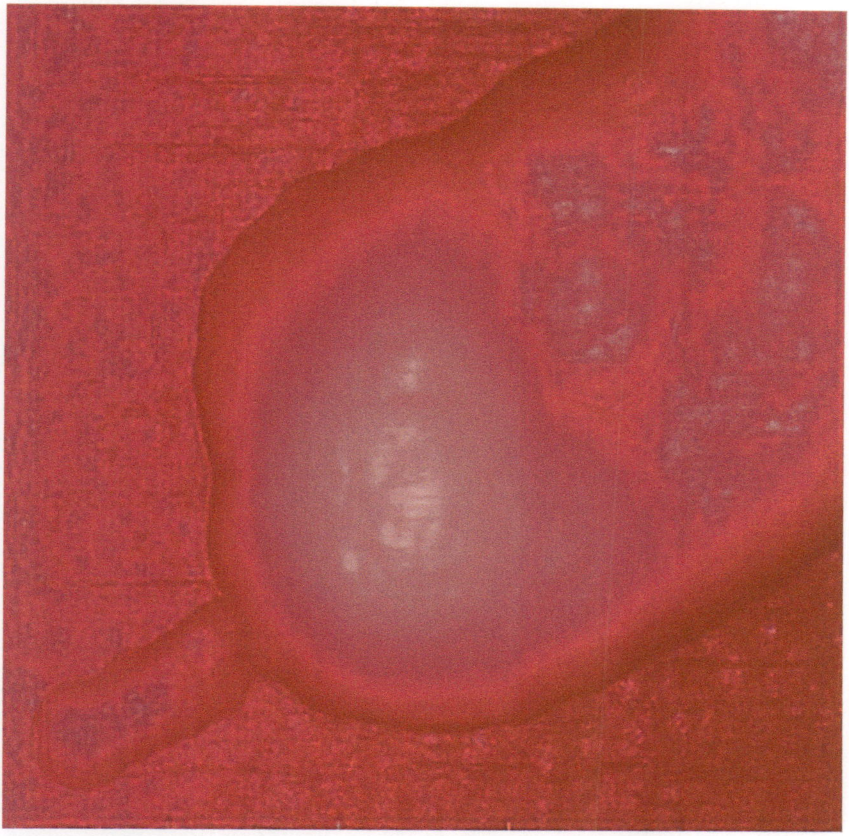

FIGURE 2.38 Height mode of atomic force microscopy shows the tubular protrusion in susceptible *Mycobacterium tuberculosis*.

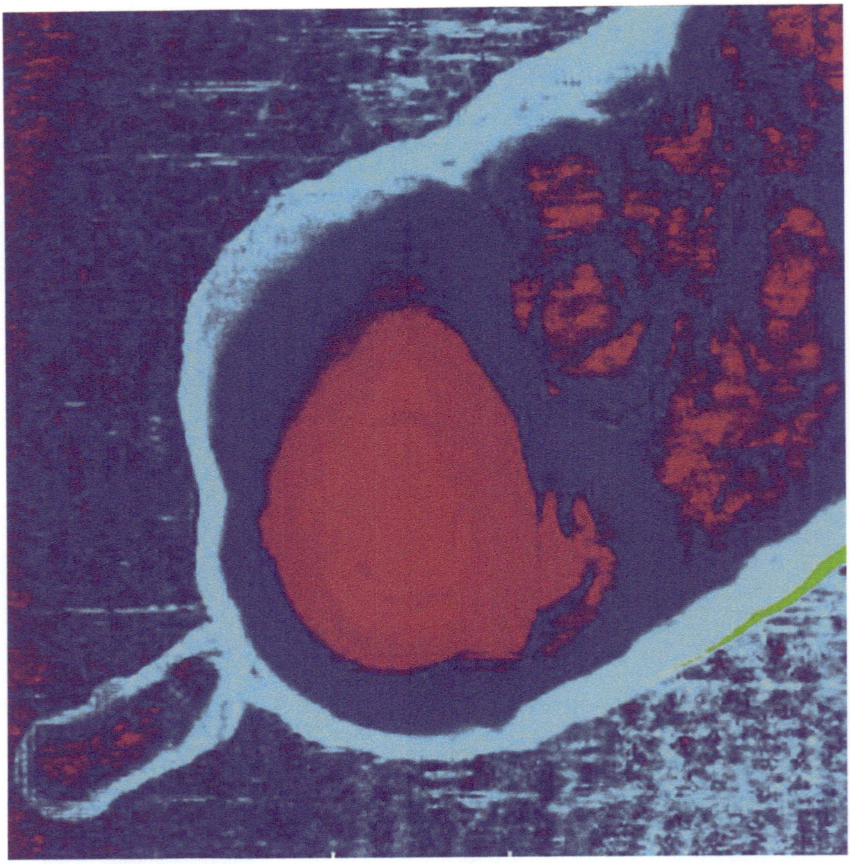

FIGURE 2.39 The tubular protrusion in susceptible *Mycobacterium tuberculosis*. These structures had a low surface roughness (1.8 ± 1.9 nm).

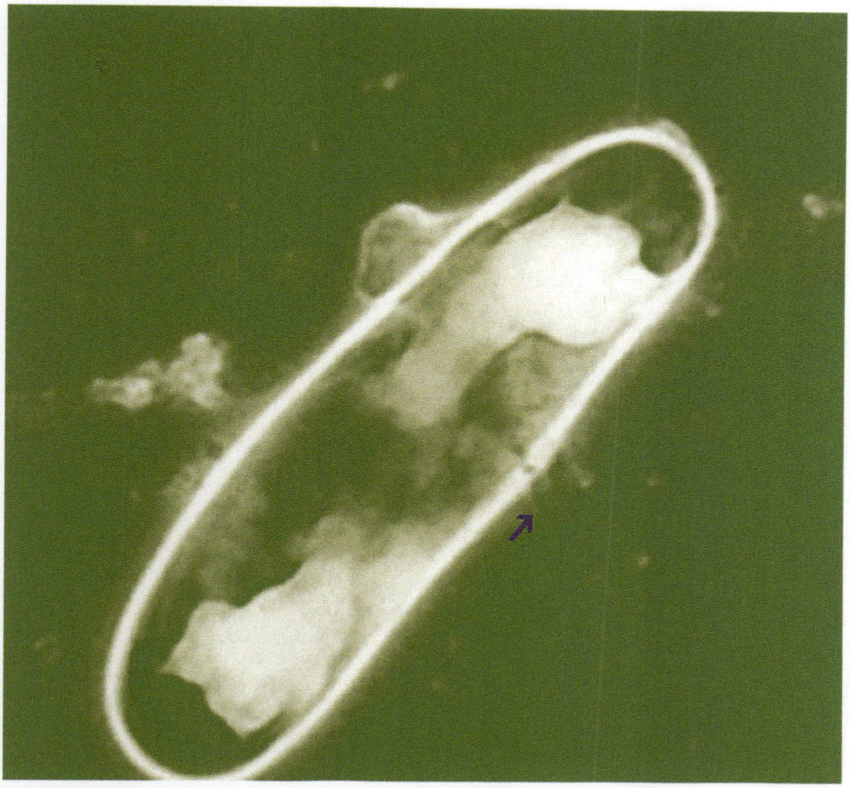

FIGURE 2.40 Transmission electron micrograph shows the tubular protrusion in susceptible *Mycobacterium tuberculosis* (*arrow*). The projection of this tubular structure can be on side walls or pole sides of *M. tuberculosis*.

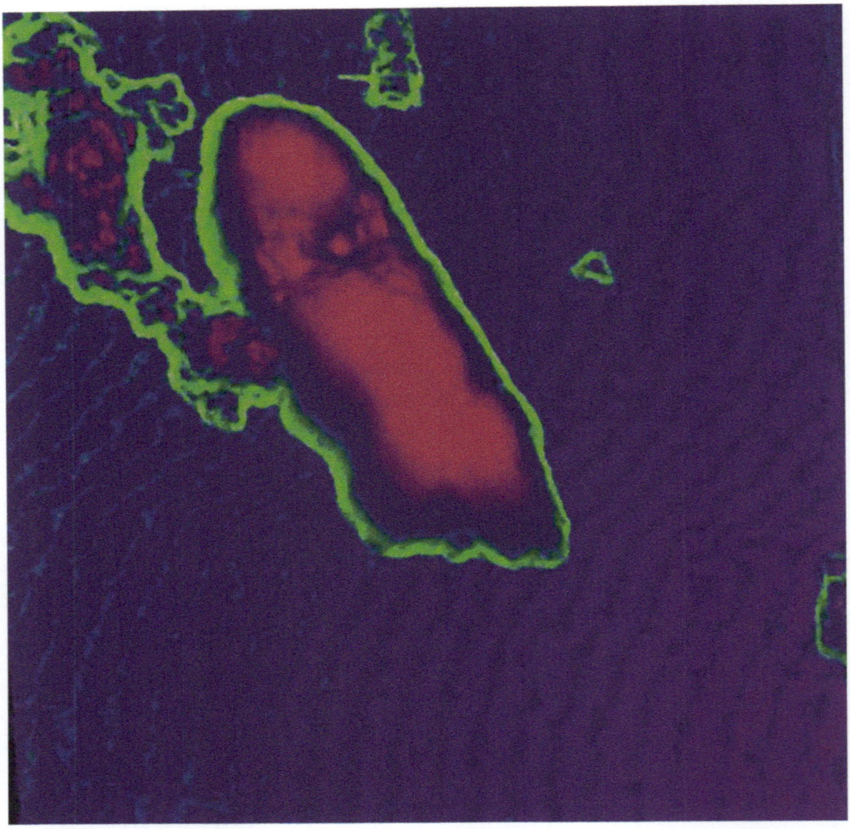

FIGURE 2.41 Discharge of material from susceptible *Mycobacterium tuberculosis*. These materials can be secretory proteins or toxic compounds.

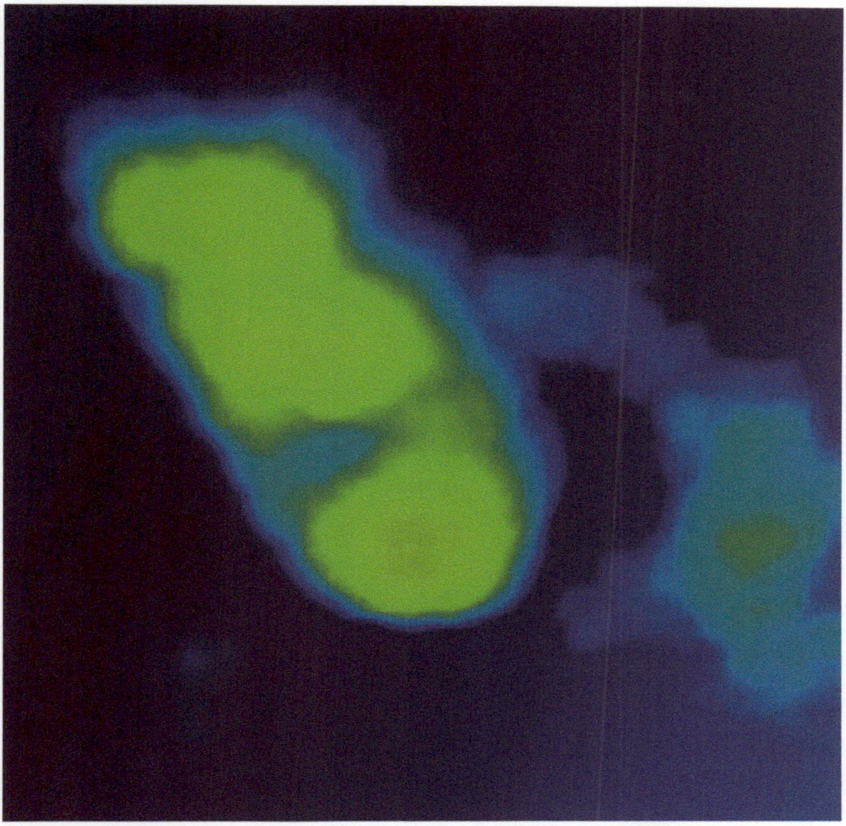

FIGURE 2.42 Discharge of material from susceptible *Mycobacterium tuberculosis* recorded by atomic force microscopy.

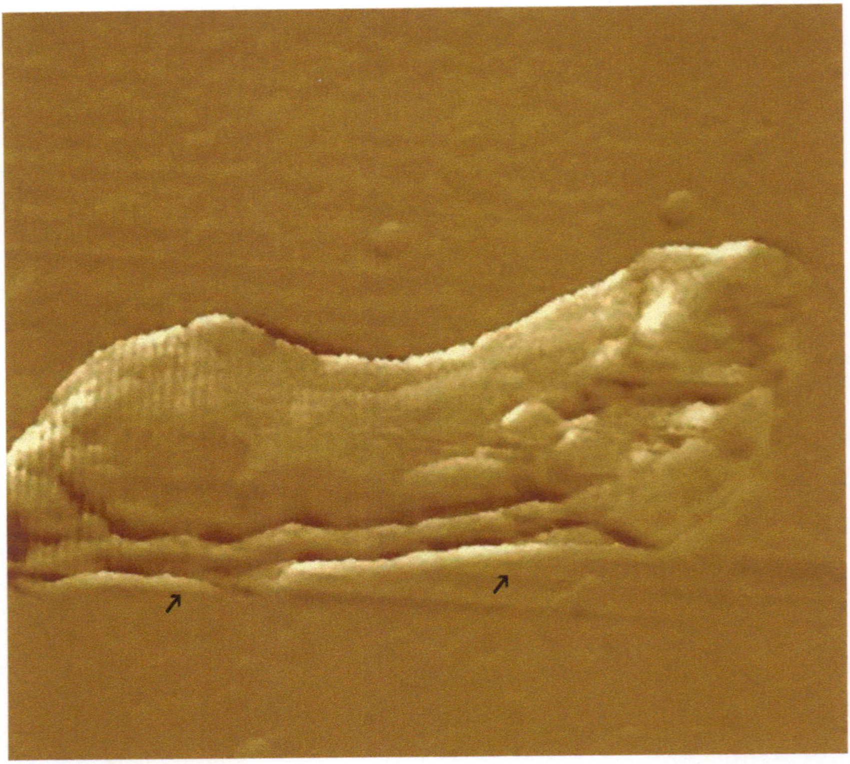

FIGURE 2.43 Totally drug-resistant tuberculosis bacilli. Sometimes the tubular protrusion (*black arrows*) is as long as the bacilli size (up to 2 μm).

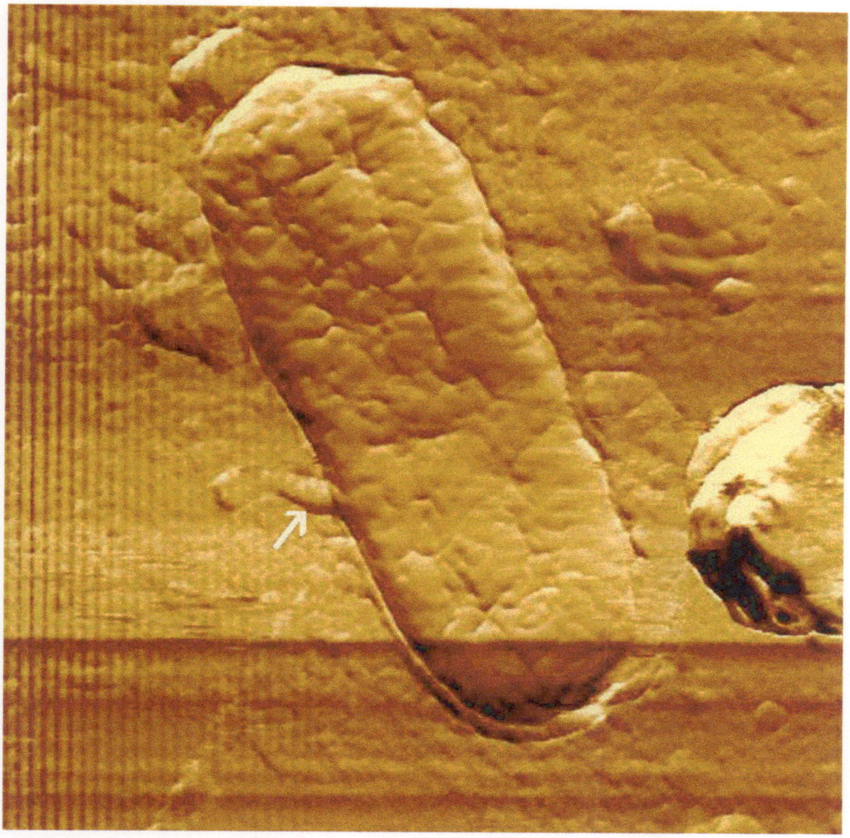

FIGURE 2.44 Atomic force microscopy recorded tubular projection in left side walls (25 nm) of *Mycobacterium tuberculosis (white arrow).*

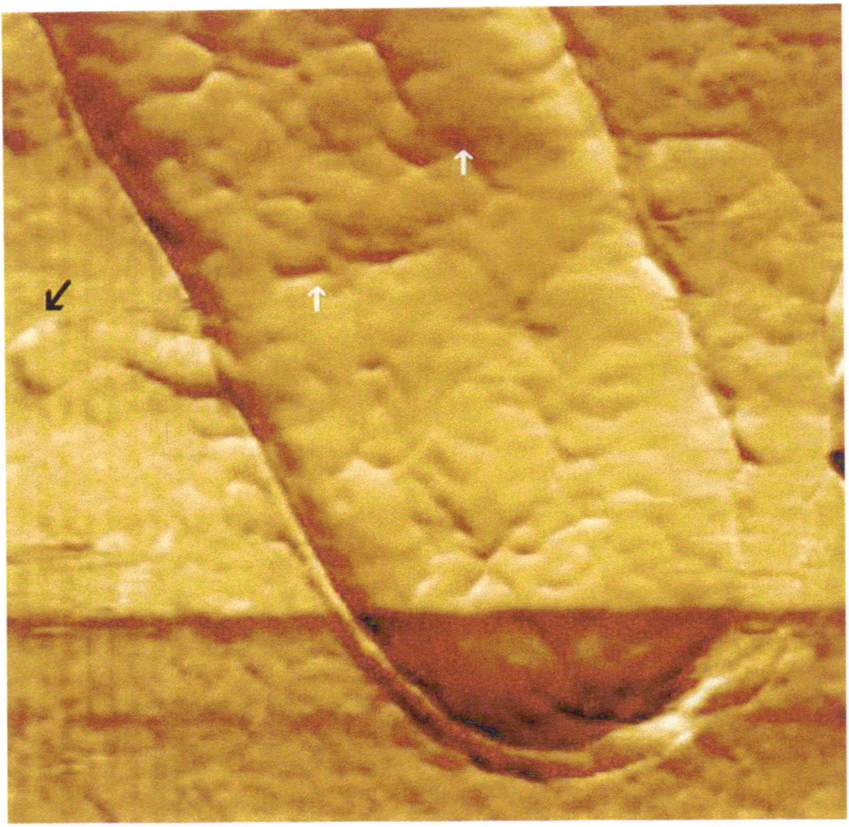

FIGURE 2.45 Friction contact mode of atomic force microscopy shows the tapering-end projection (*black arrow*). On the surface of the cell the porins or pores are visible (*white arrows*).

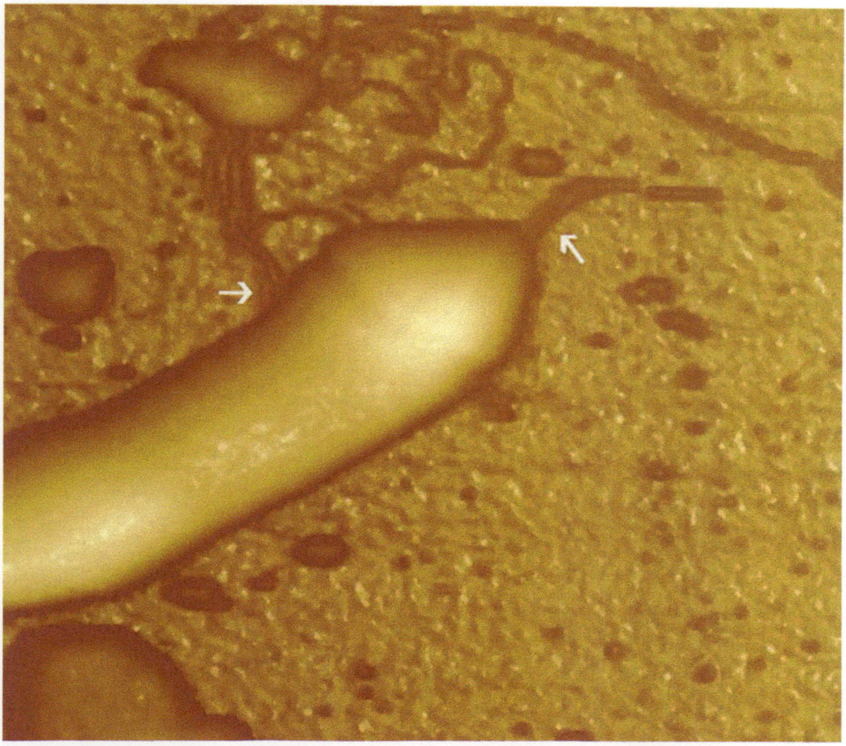

FIGURE 2.46 Pili in *Mycobacterium tuberculosis.* Pili are proteinaceous polymeric structures generally composed of a major repeating subunit or pilin and, in some cases, a minor tip-associated adhesion subunit. They are straight or flexible filaments, 2–10 nm wide and 2–3 μm long (*white arrows*).

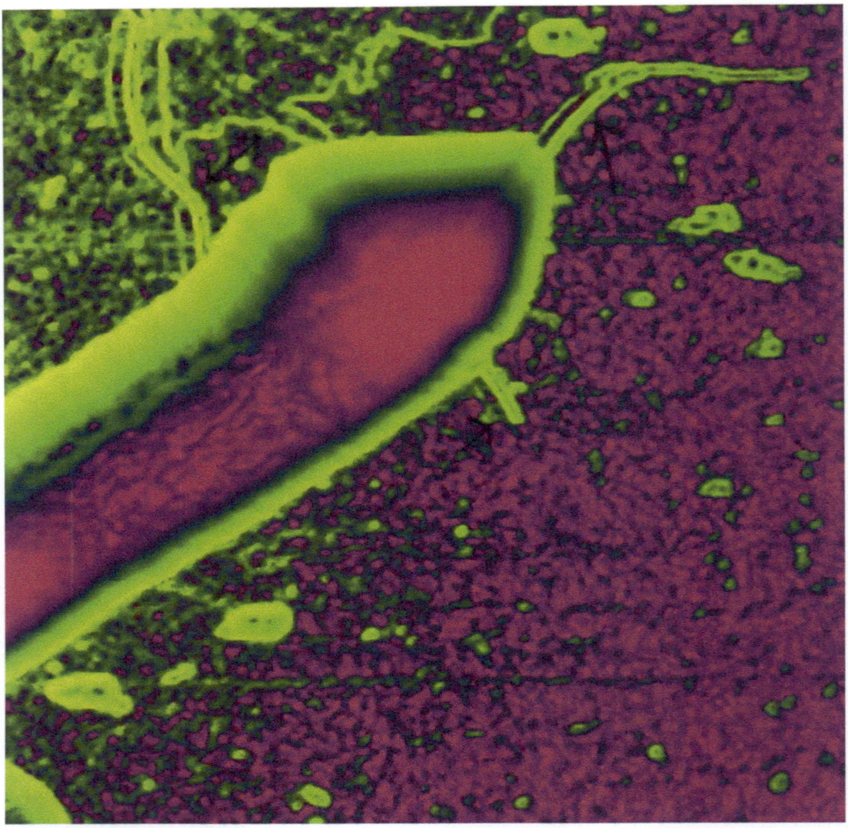

FIGURE 2.47 Pili in *Mycobacterium tuberculosis* recorded by atomic force microscopy (*black arrows*).

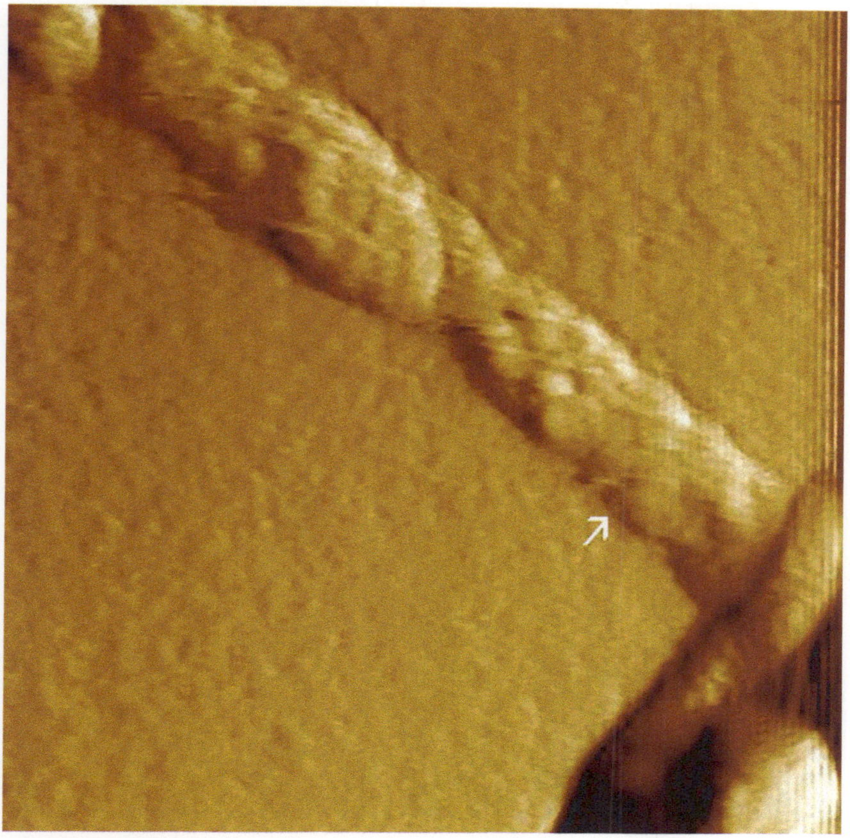

FIGURE 2.48 Contact mode of atomic force microscopy shows pili in *Mycobacterium tuberculosis.* Pili can protrude from different sides of the tuberculosis bacilli surface (*white arrow*). Pili can be in coiled-coin aggregate fibers or they can be as a single fiber tube-type structure that protrude from the head, tail, or side poles of the bacilli.

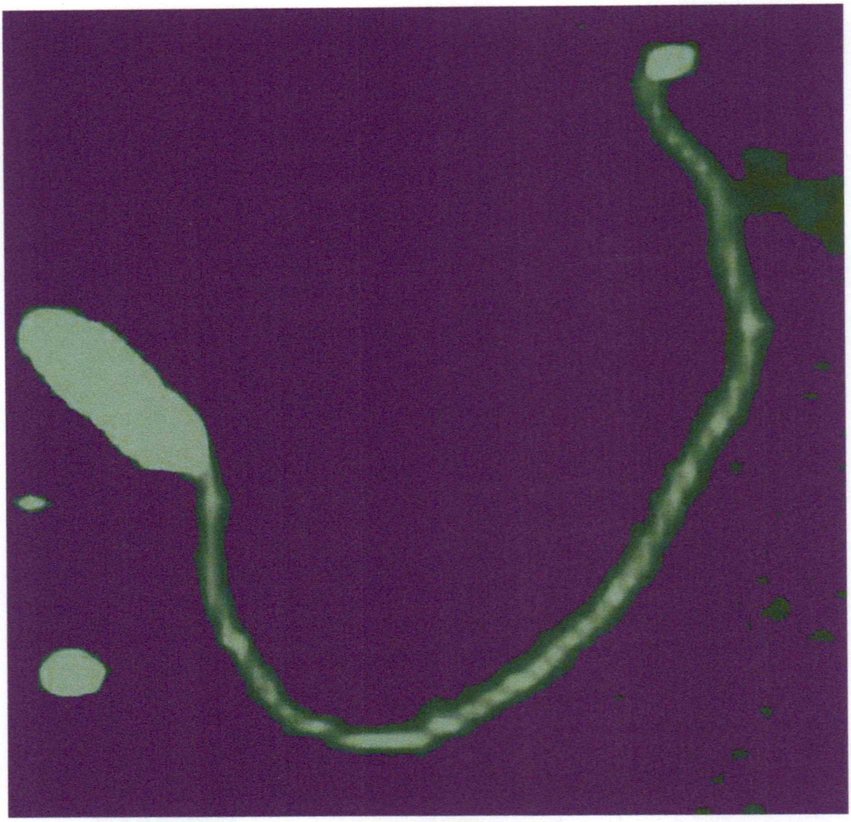

FIGURE 2.49 Pili in *Mycobacterium tuberculosis*. Sometimes in totally drug-resistant tuberculosis bacilli the size of the pili reaches to 30 ± 5 nm with an average of 3 μm in diameter.

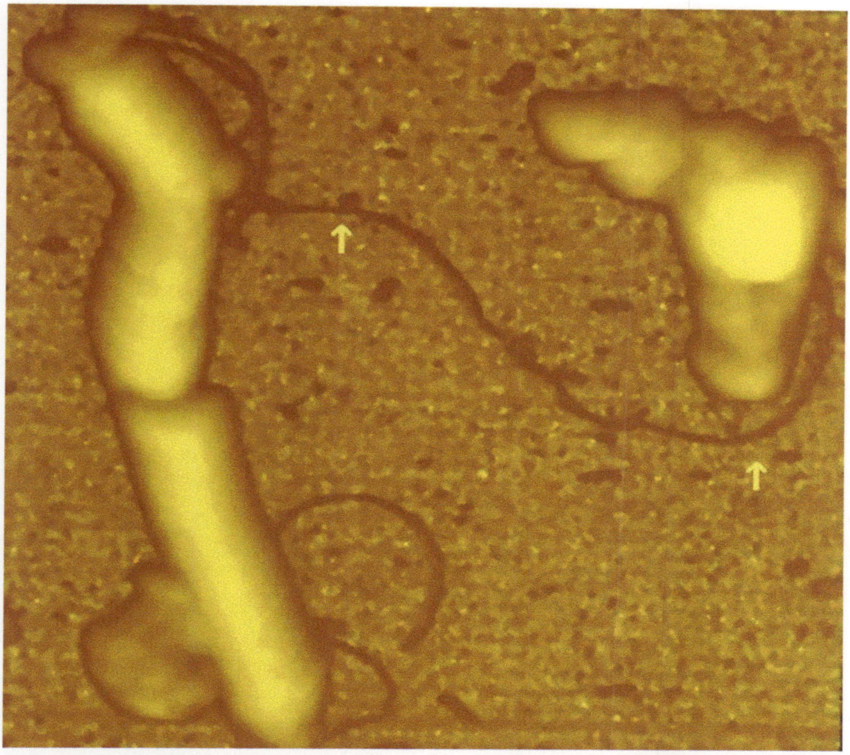

FIGURE 2.50 Cell-to-cell communication by pili (*yellow arrows*). *Mycobacterium tuberculosis* can communicate with each other using pili as connectors. In general, pili have two basic functions: gene transfer and attachment to surfaces. Thus far the assembly and adhesive functions were suggested for *Mycobacterium* pili, but their exact relevance with regard to biological and immunological function is yet to be discovered.

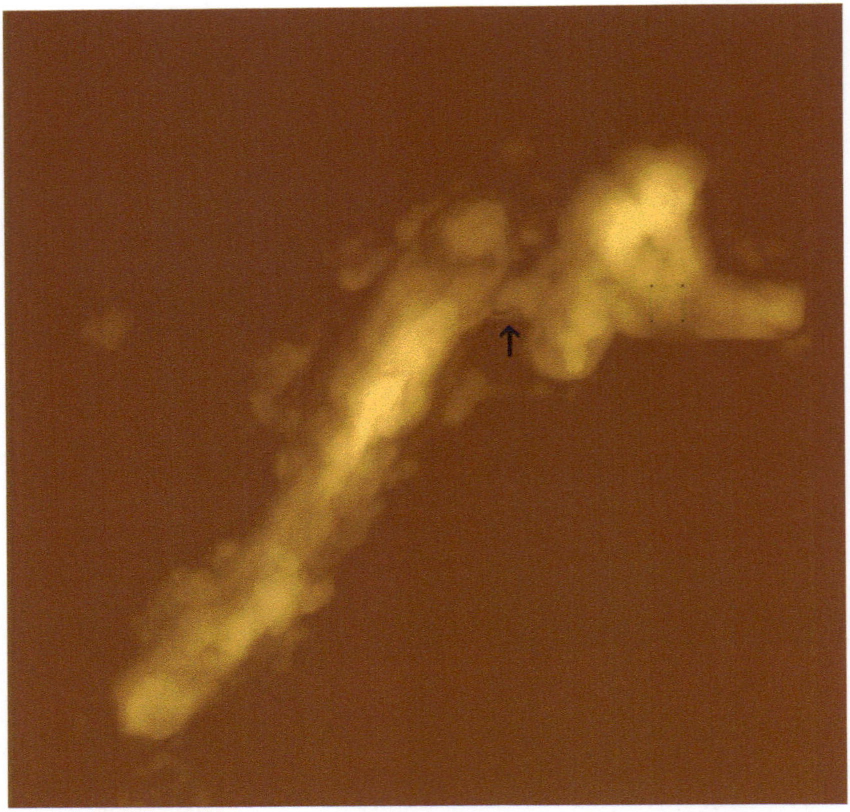

FIGURE 2.51 Cell-to-cell communication in *Mycobacterium tuberculosis* recorded by atomic force microscopy. Using a high-resolution technique it becomes clear that *M. tuberculosis* are communicating with each other (*black arrow*). However, it is not known if they can also exchange the genetic material.

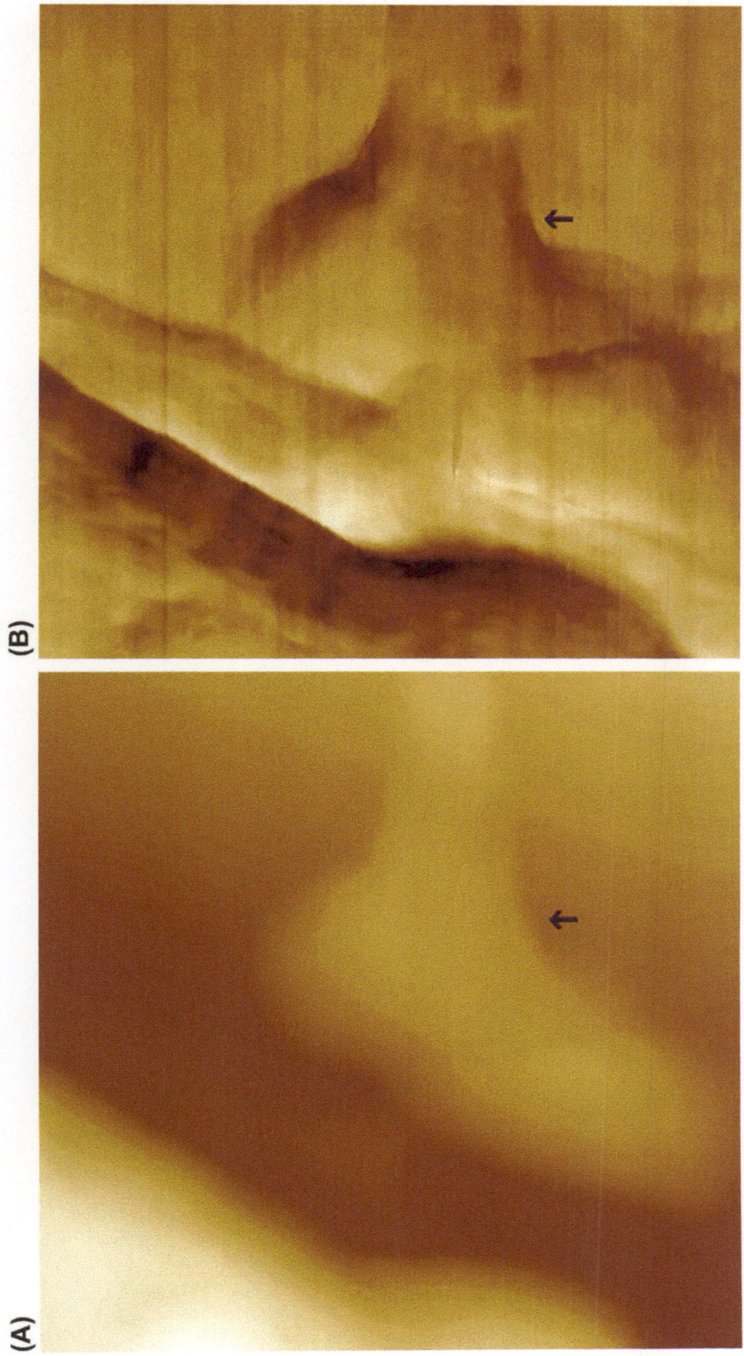

FIGURE 2.52 Cross-section of cell-to-cell communication in *Mycobacterium tuberculosis* by pili like structure (*black arrows*) under (A) height and (B) friction mode of atomic force microscopy. Communication and/or exchange of genetic material might open new intriguing gates that could possibly challenge the impact of future mycobacteriology research.

Chapter 3

Diversity in Cell Shape of *Mycobacterium tuberculosis*

Under the light microscope, tubercle bacilli typically appear as a straight or slightly curved rod (Ziehl, 1882). The dimensions of bacilli varied from 1 to 10 μm in length (usually 3–5 μm) and 0.2–0.6 μm in width (Whitman et al., 2012). They are metabolically flexible and are capable of adapting to any environmental changes throughout the course of infection (Ramakrishnan et al., 1972; Niederweis, 2008). The success of *Mycobacterium tuberculosis* as a pathogen can be attributed to its extraordinary capacity to change its shape under stress conditions (Edson, 1951; McCune and Tompsett, 1956; Mattman, 1970; Ramakrishnan et al., 1972; Nyka, 1974; Takahashi, 1979; Khomenko et al., 1983; Wayne, 1976, 1981, 1994; Wayne and Sohaskey, 2001; Niederweis, 2008; Cook et al., 2009). During the last 2 decades the possibility of morphological variations in tubercle bacilli was suggested by many investigators (Malassez and Vignal, 1883; Nocard and Roux, 1887; Metschnikoff, 1888; Lubarsch, 1899; Fischel, 1893; Vera and Rettger, 1940). They showed under unfavorable conditions (ie, a limited food supply or oxygen deprivation) that *Mycobacterium* assumed a swollen appearance without forming the vacuolar or globoid bodies (Vera and Rettger, 1940). However, these early reports were based on stained preparations and were subjected to severe criticism (Porter and Yegian, 1945). Today with advances in microscopic techniques (ie, transmission electron microscopy, scanning electron microscopy, and atomic force microcopy) almost all investigators have agreed that the Koch bacillus does not always manifest itself in the classical rod shape. Among the factors that can influence the shape of *M. tuberculosis* are growth conditions, ages of the cultures, nutritional quality, physical conditions (hypoxia, temperature, pH, salinity), nutrient deprivation, and various exogenous stress conditions (Vera and Rettger, 1939; Wheate, 1951; Bongiorno, 1953; Nakamura, 1958; Khomenko et al., 1983; Smeulders et al., 1999; Honer et al., 2001; Young et al., 2005; Anuchin et al., 2009; Velayati et al., 2009a,b, 2011, 2016; Farnia et al., 2010; Singh et al., 2010; Shleeva et al., 2002, 2011; Markova et al., 2012; Sieniawska et al., 2015). Unfortunately, in most of cases we do not know if shape per se is beneficial because the knowledge of *M. tuberculosis* physiology during this process has been limited by the slow growth of the bacterium in the laboratory. We recently designed an experi-

Atlas of *Mycobacterium tuberculosis*.

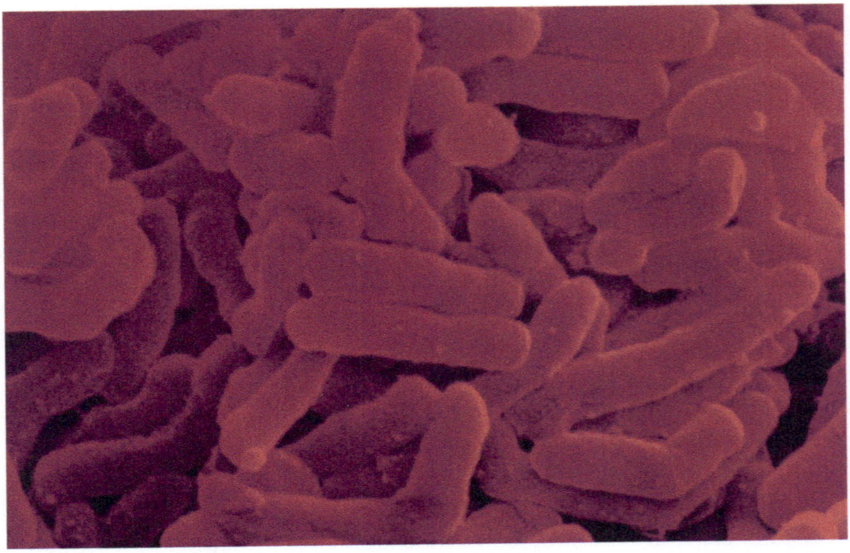

FIGURE 3.1 Scanning electron microscopy (SEM) of *Mycobacterium tuberculosis* at the exponential phase of growth. Cells inactivated in 1 mL of solution containing 2% paraformaldehyde and 2.5% glutaraldehyde in 0.1 M sodium cacodylate buffer (pH 7.4). Then it was treated with 2% osmium tetroxide in 0.1 M cacodylate buffer for 2 h at room temperature. Dehydration was followed using a sequential grade of ethanol (50%, 70%, 95%, and 100%). Dried samples were gold-sputter–coated and imaged under SEM.

mental setup to study *M. tuberculosis* under different growth conditions. We classified the morphological variation of *M. tuberculosis* into two categories: those that are frequently seen at the exponential phase of growth (rod, V, or Y shape, branched or buds) and those that are seen occasionally under stress or environmental conditions (round, oval, ultravirus, spore-like, and cell wall defiant or L-forms; Velayati et al., 2009a,b,c, 2012a,b; Farnia et al., 2010; Velayati and Farnia, 2012a,b; Velayati et al., 2016) (Figs. 3.1–3.29).

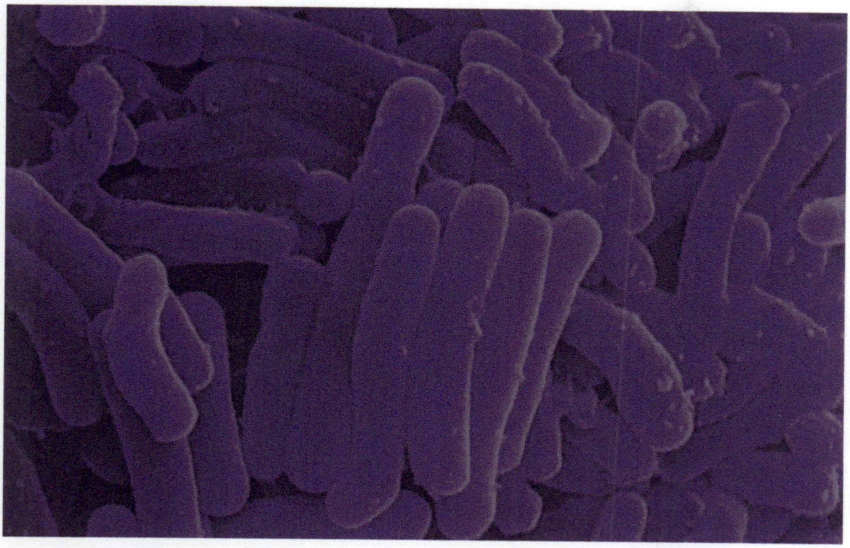

FIGURE 3.2 Most of the *Mycobacterium tuberculosis* population are slender, rod-shaped bacilli. The dimensions of the bacilli have been reported to be 1–10 µm in length (usually 3–5 µm) and 0.2–0.6 µm in width.

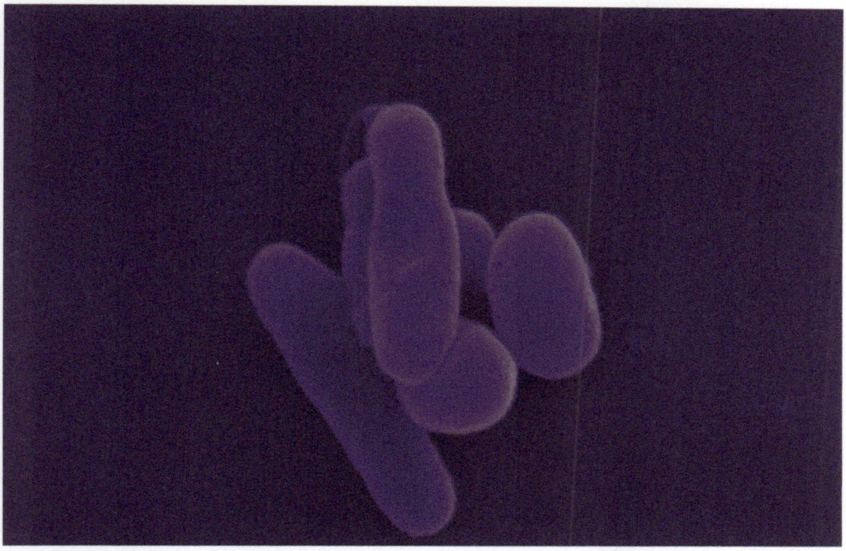

FIGURE 3.3 The most classical form of *Mycobacterium tuberculosis* is the slender rod shape that is seen in stained smears, but the bacilli can also be in round form. The number and size of round bacilli are variable in different growth conditions.

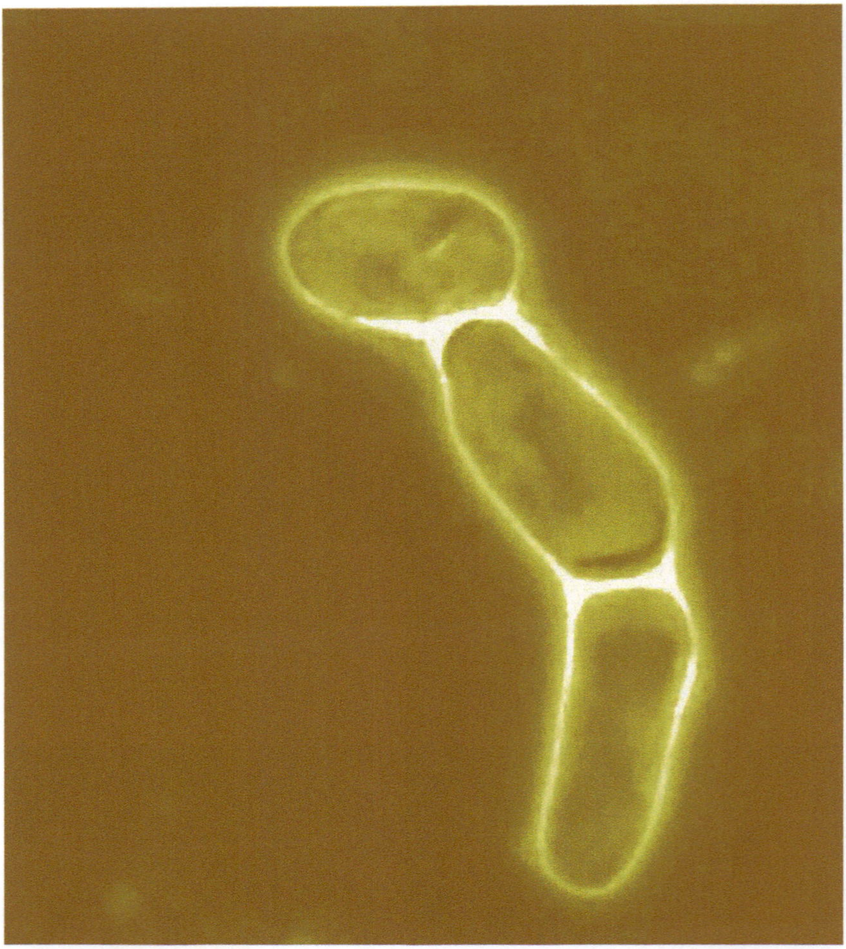

FIGURE 3.4 *Mycobacterium tuberculosis* were negatively stained using a 3% solution of uranyl acetate and examined under transmission electron microscopy. The population of bacilli has a different shape and size in the exponential phase of growth.

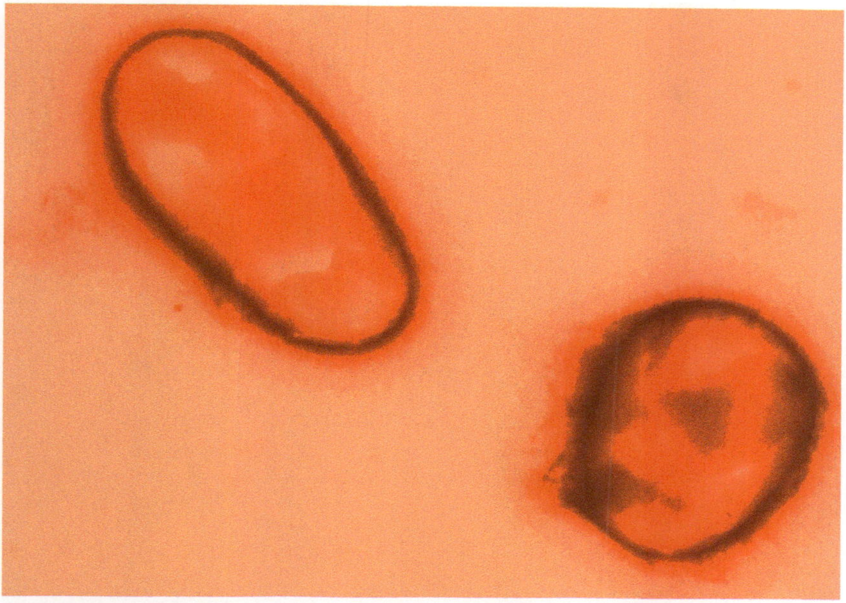

FIGURE 3.5 The size of round *Mycobacterium tuberculosis* may be variable from 0.9 to 0.3 μm depending on their growth conditions.

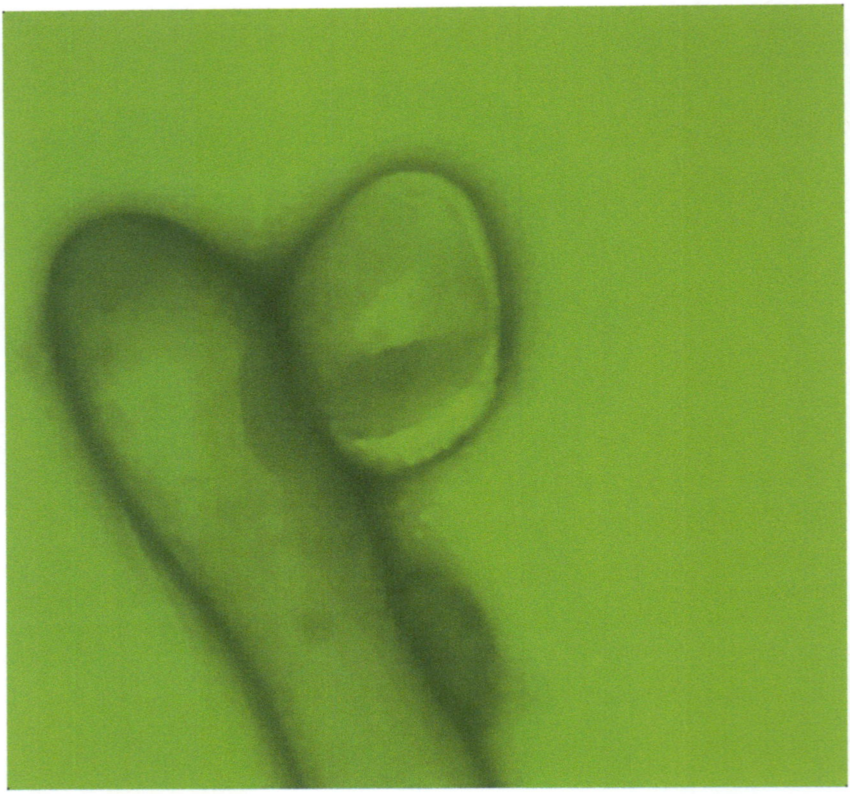

FIGURE 3.6 The round-shaped *Mycobacterium tuberculosis* can be formed by asymmetrical or budding type of cell divisions.

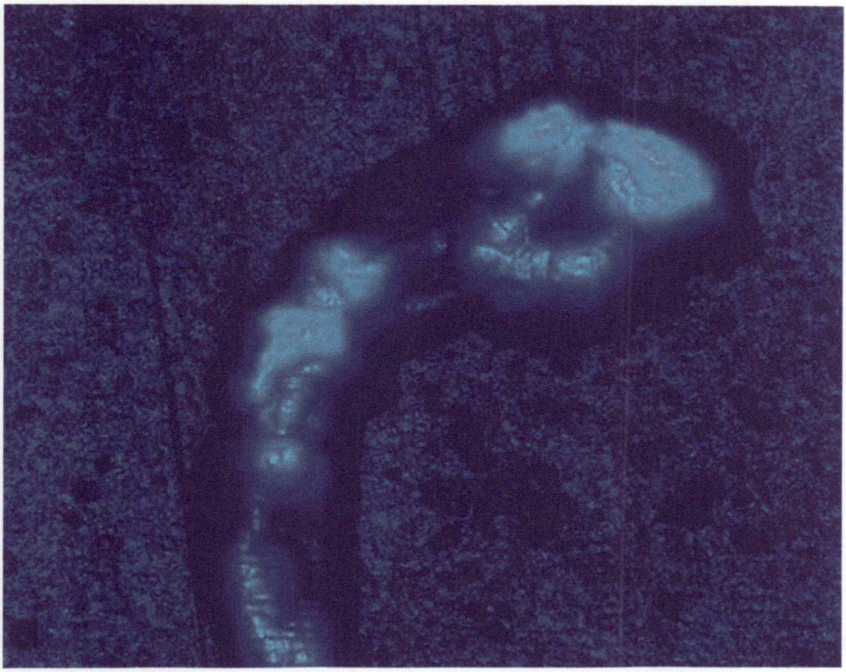

FIGURE 3.7 The round-shaped *Mycobacterium tuberculosis* can be formed by an asymmetrical type of cell. The atomic force microscopy image shows the internal details of round-shaped bacilli. Similar to its rod-like mother cells, the round cells also consist of dense nuclear masses within granular cytoplasm.

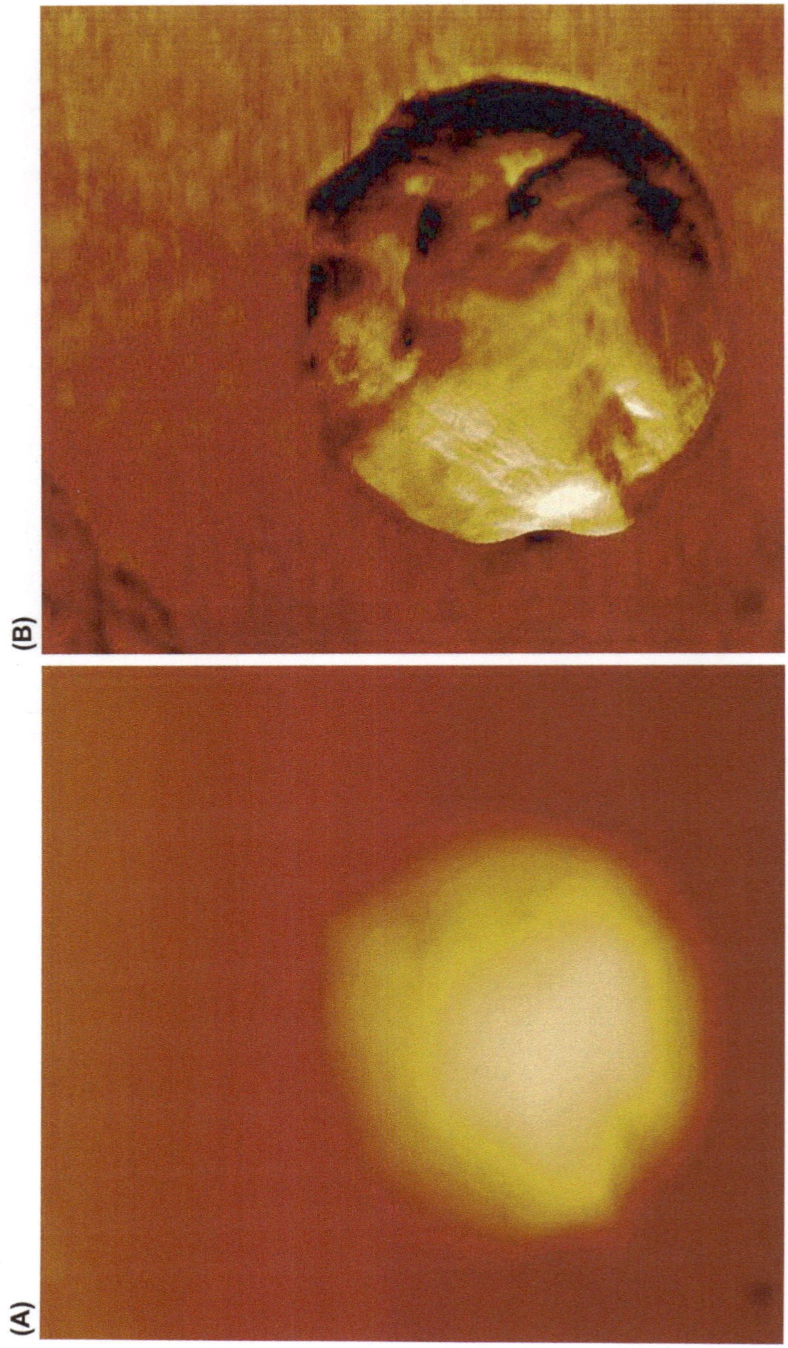

FIGURE 3.8 The round-shaped *Mycobacterium tuberculosis* recorded by the (A) height and (B) deflection mode of atomic force microscopy. The number of round-shaped cells is low in susceptible tuberculosis populations (1–2%) whereas their number increases considerably in extensive and extreme drug-resistant strains (10–15%). These morphological changes may help them to escape the immune response and/or overcome the inhibitory effect of most antibiotics.

FIGURE 3.9 *Mycobacterium tuberculosis* is sometimes oval in shape. The ovoid forms of bacilli are characterized by low metabolic activity and elevated resistance to antibiotics.

FIGURE 3.10 Image of ovoid-shaped *Mycobacterium tuberculosis* recorded by the deflection mode of atomic force microscopy. To image tubercle bacilli on silicon plates, the surface was charged with polyclonal rabbit anti-mycobacterium antibody.

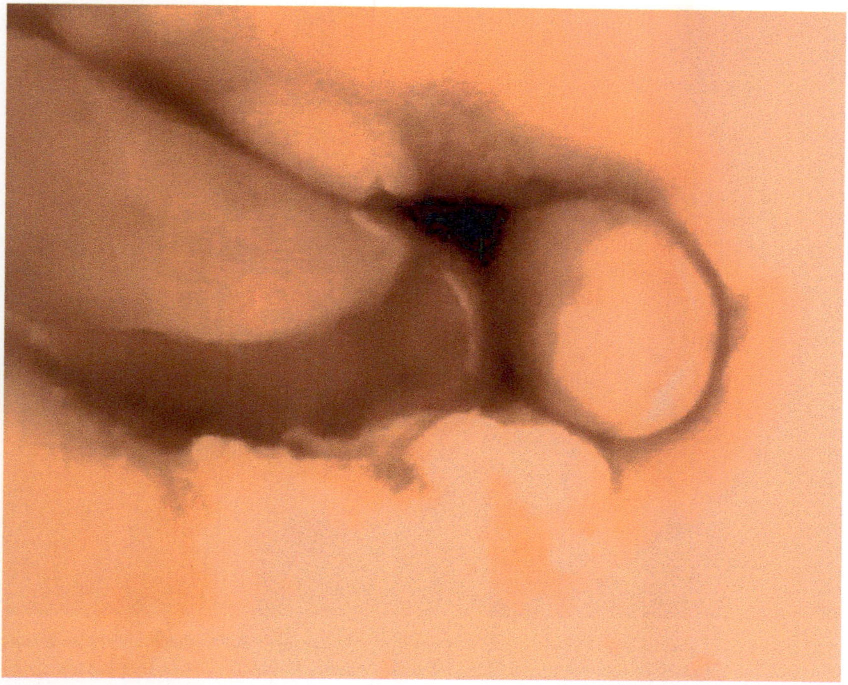

FIGURE 3.11 Transmission electron microscopy shows the round- and rod-shaped *Mycobacterium tuberculosis*. It was recently suggested that shapes and not the size are the dominant factor in determining whether particles were phagocytized or not.

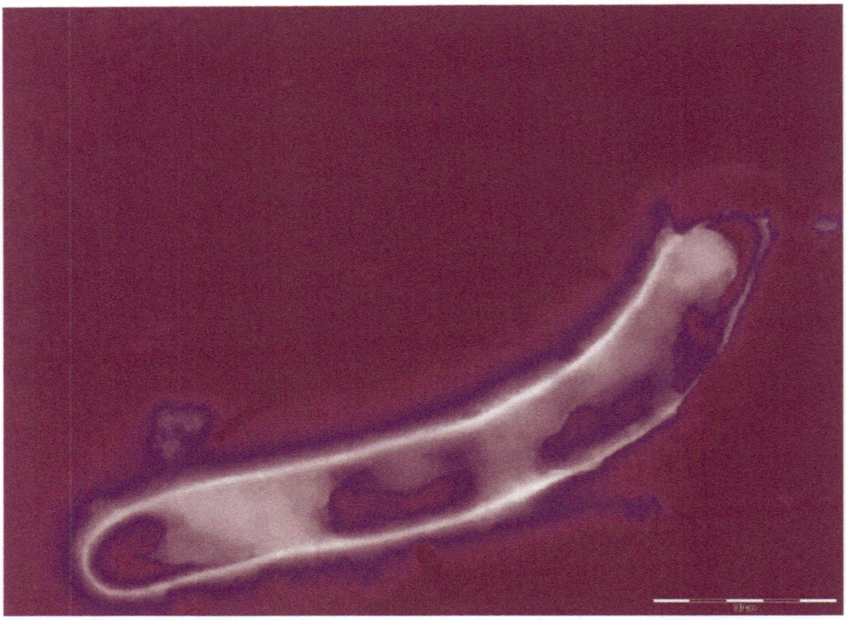

FIGURE 3.12 Transmission electron microscopy shows the elongated rod-shaped *Mycobacterium tuberculosis*. The size of the rod-shaped *M. tuberculosis* varies from 4.3×0.4 to 1.0×0.2 μm.

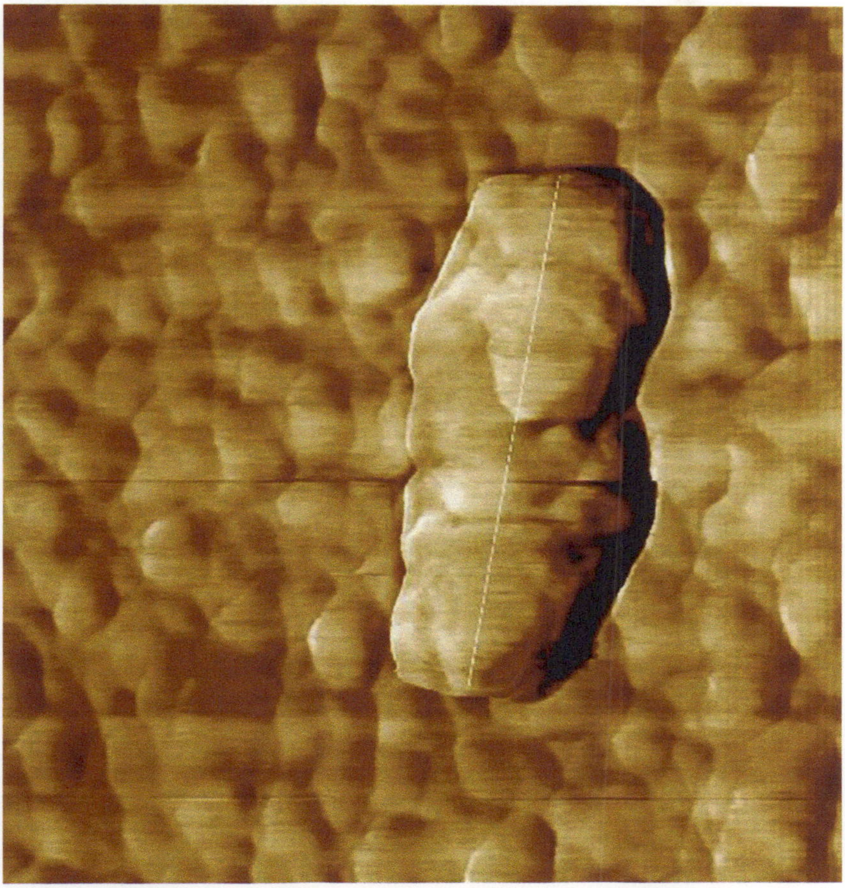

FIGURE 3.13 *Mycobacterium tuberculosis* under stress conditions reduces its size. Atomic force microscopy shows the rod-shaped bacilli under stress or harsh conditions (ie, excess antibiotics) with size reduction (from 0.9 × 0.2 to 0.5 × 0.2 μm).

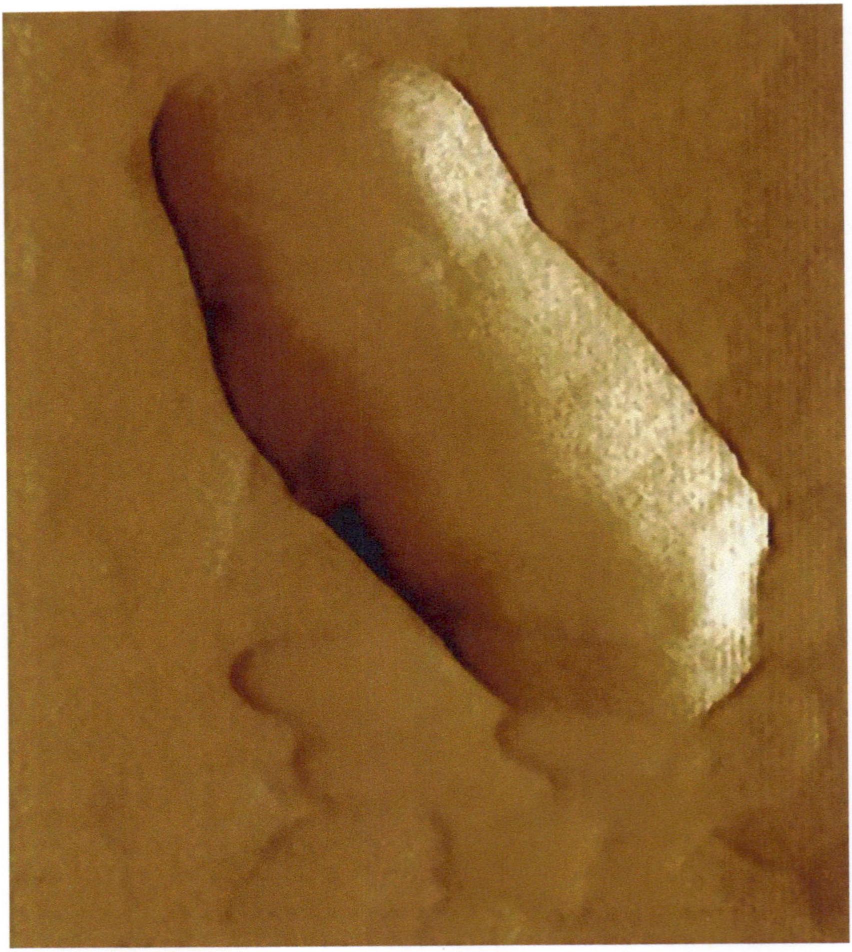

FIGURE 3.14 *Mycobacterium tuberculosis* has a smooth cell surface in the absence of harsh conditions. The average surface roughness on 200×200 nm is approximately 8.0±2.5 nm with a cell wall thickness of 12–15 nm.

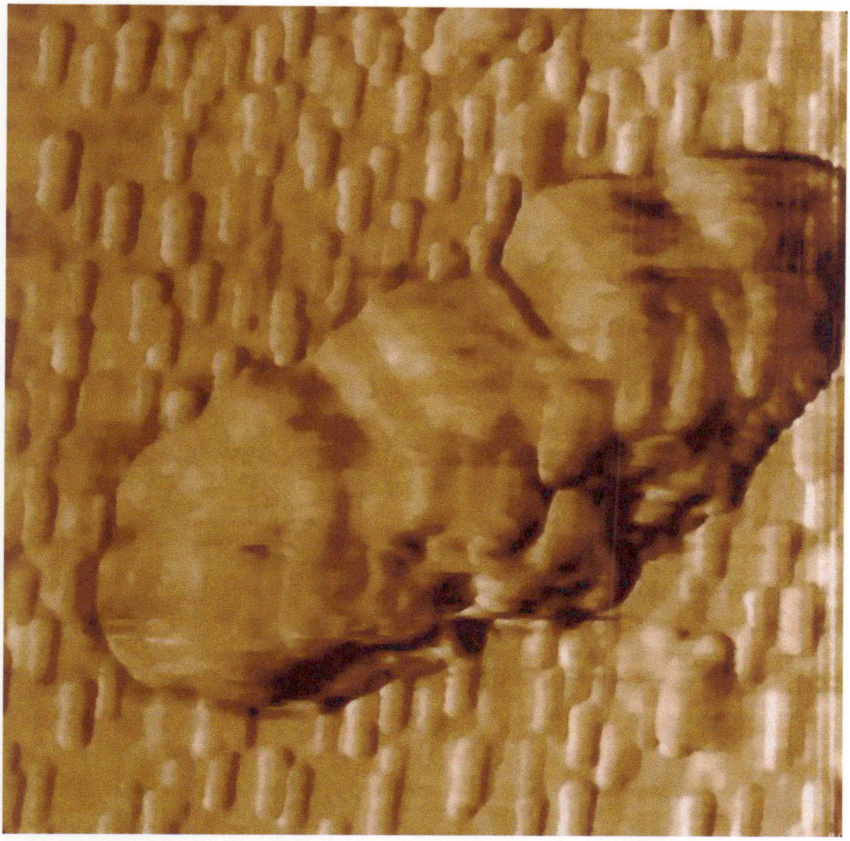

FIGURE 3.15 The surface of *Mycobacterium tuberculosis* under anaerobic cultures (30–50 days) develops a dense and homogenous layer that covers the bacilli.

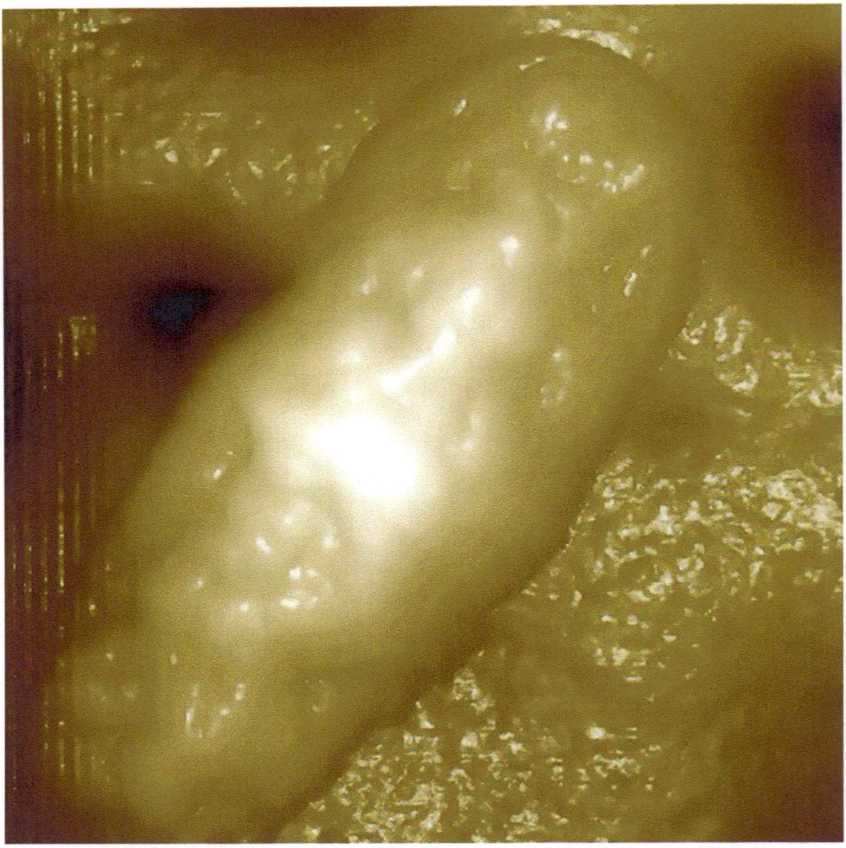

FIGURE 3.16 The surface of totally drug-resistant *Mycobacterium tuberculosis* recorded by atomic force microscopy. The average surface roughness increases and reaches ~21.5 ± 1.8 nm.

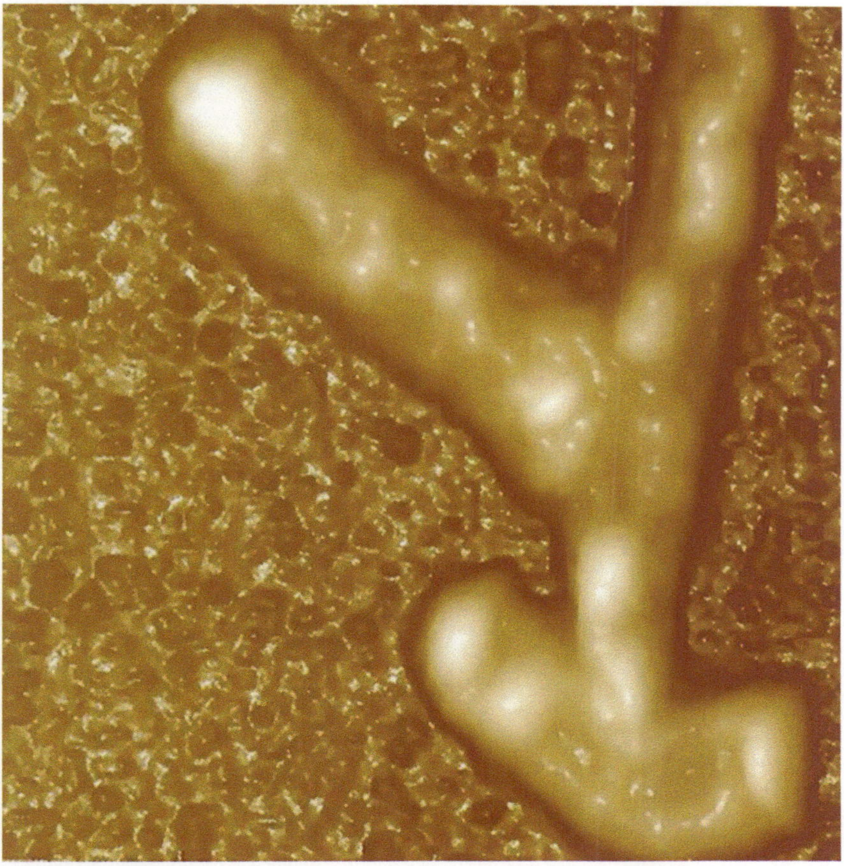

FIGURE 3.17 The V-shaped *Mycobacterium tuberculosis* recorded by the height mode of atomic force microscopy. Formally, the V forms arise by any of three methods: (1) germinating of adjacent coccoid elements, (2) subpolar germination (budding) of rods, and (3) snapping postfission movements.

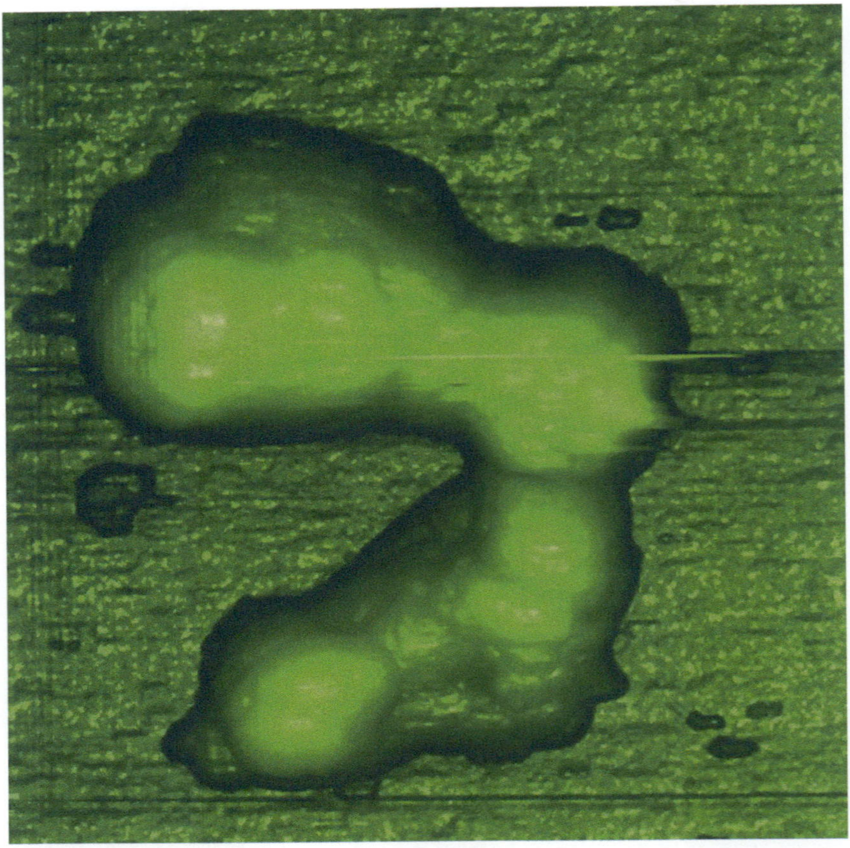

FIGURE 3.18 The V-shaped *Mycobacterium tuberculosis*. In tubercle bacilli, during septum formation the plasma membrane and inner cell wall grow inward but the outer cell wall layer remains intact. Upon completion of septum formation with a cross-wall, the inner layer may continue to grow and thus exert pressure upon the outer cell wall layer. The outer layer eventually ruptures first on one side of the cell, and the two daughter cells bend in on the side where the outer layer is still intact, forming a V shape.

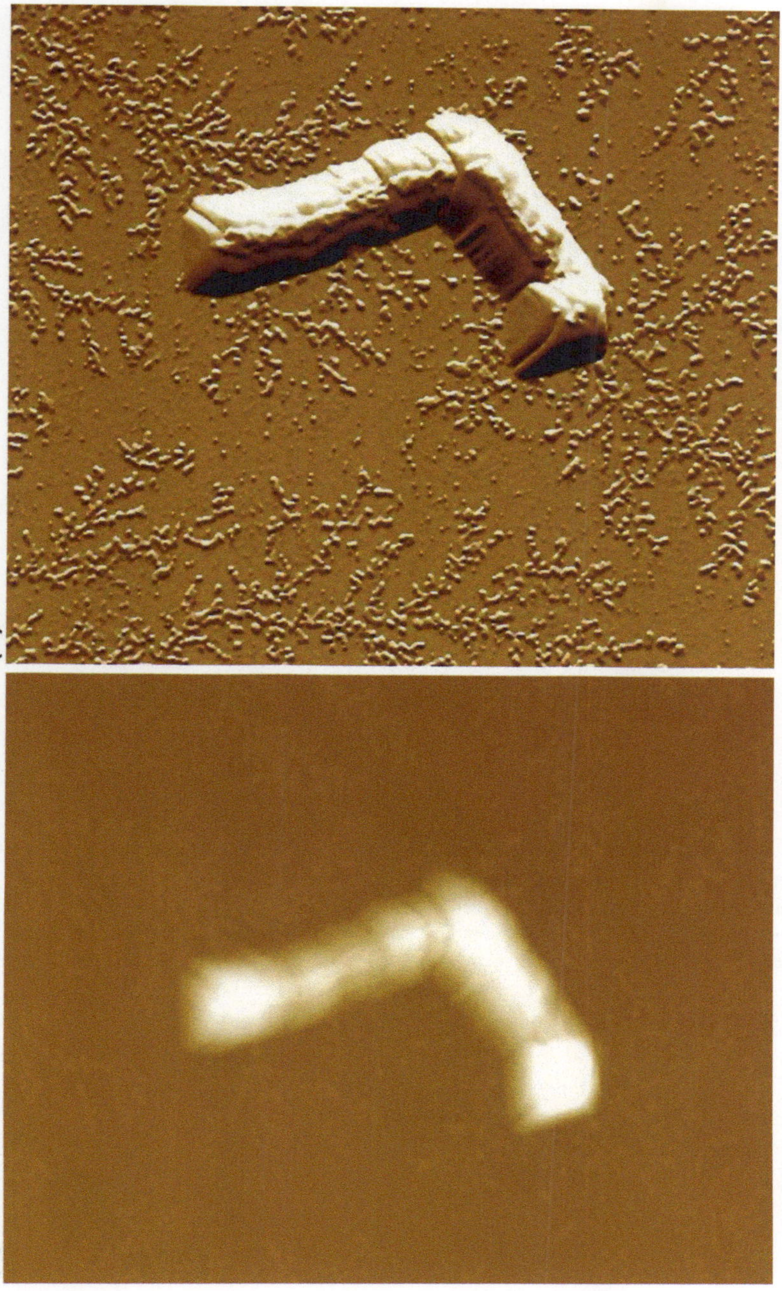

FIGURE 3.19 The V-shaped *Mycobacterium tuberculosis* recorded by (A) height and (B) deflection mode of atomic force microscopy.

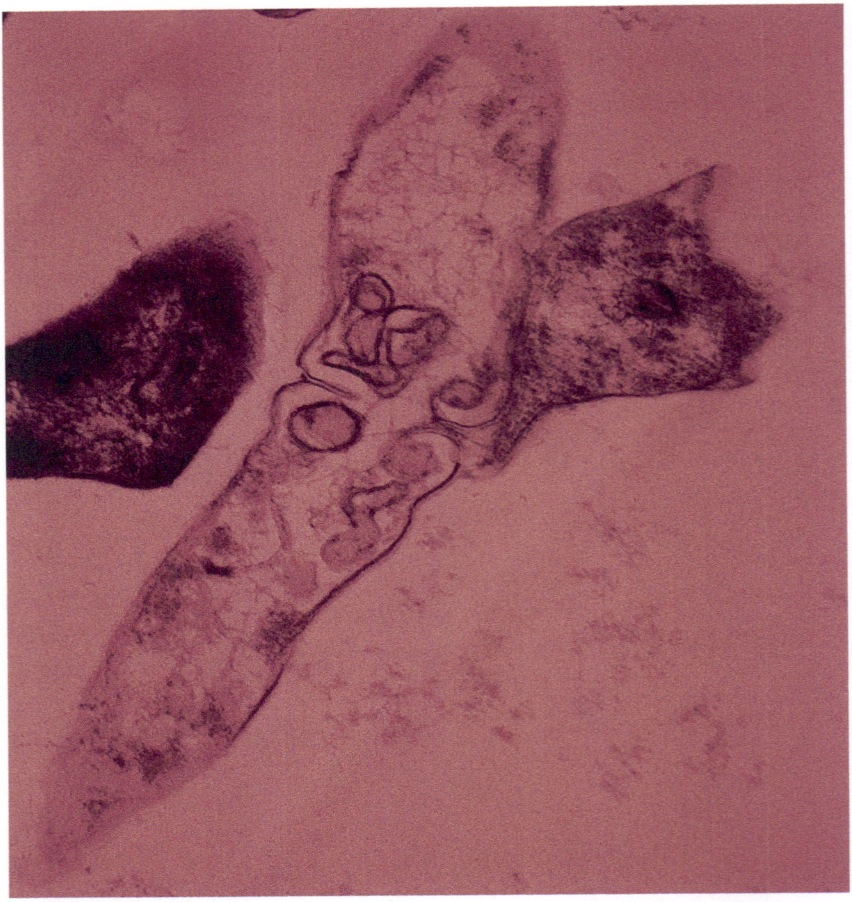

FIGURE 3.20 The heat-killed *Mycobacterium tuberculosis* in exponential phase of growth. The Y-shaped tuberculosis bacilli are also present in susceptible populations. It is believed that branching in *Mycobacterium* is similar to Actinomycetes branching in general—transient and probably a means to allow for simple, vegetative division.

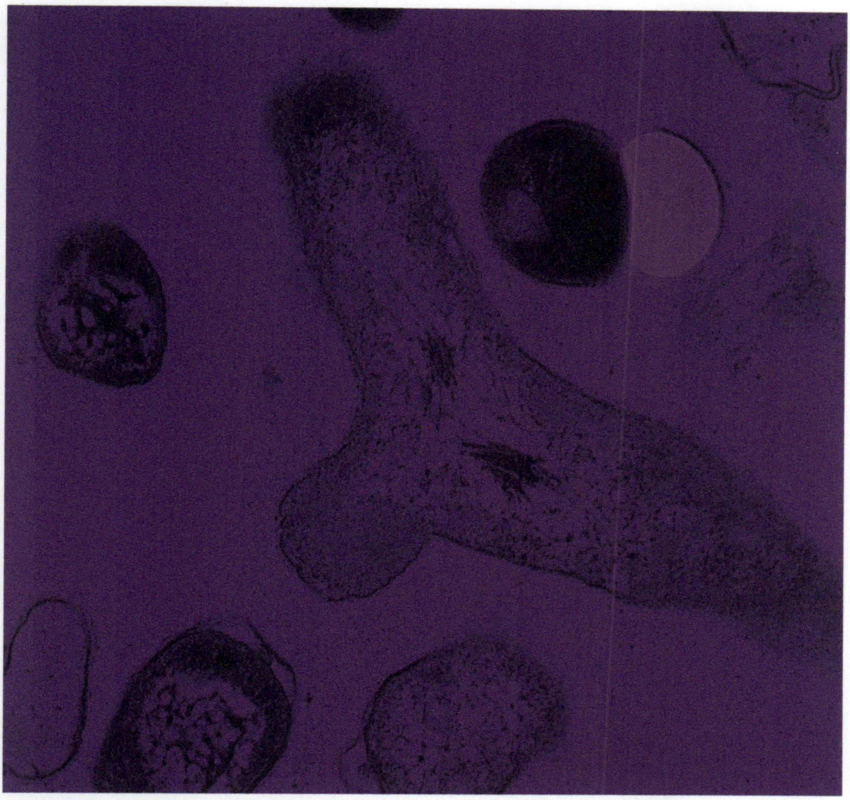

FIGURE 3.21 The Y-shaped *Mycobacterium tuberculosis* under transmission electron microscopy.

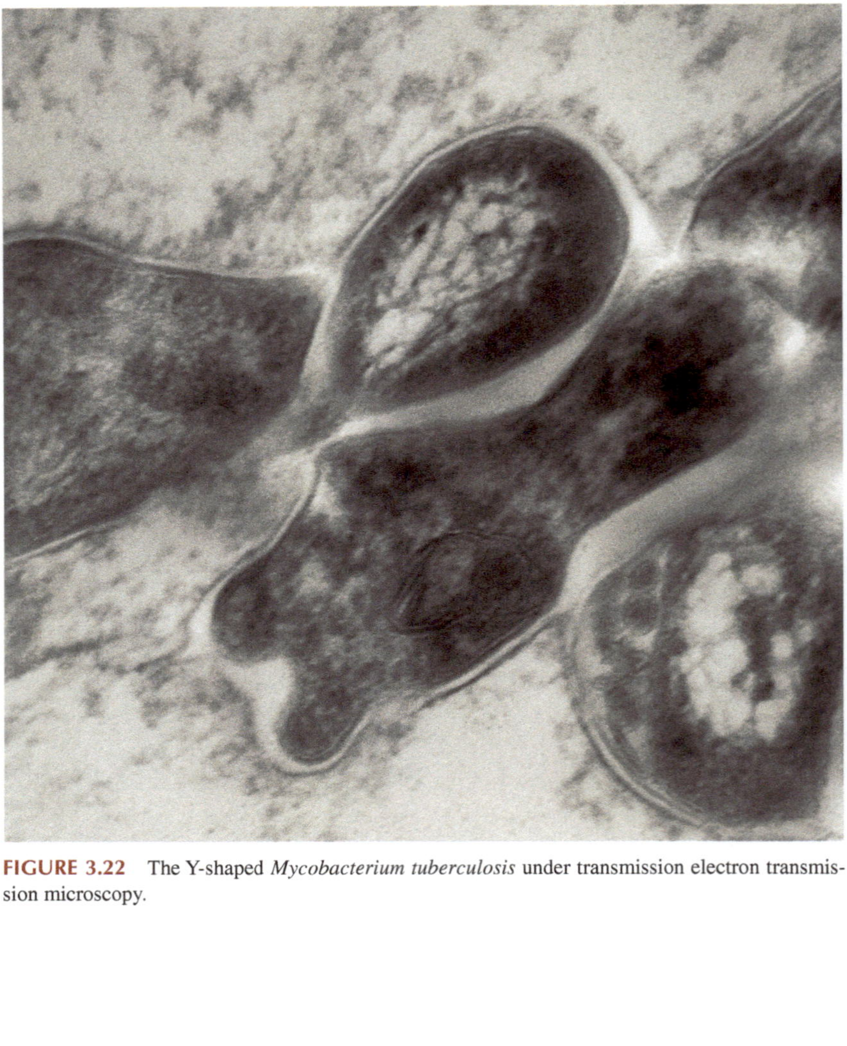

FIGURE 3.22 The Y-shaped *Mycobacterium tuberculosis* under transmission electron transmission microscopy.

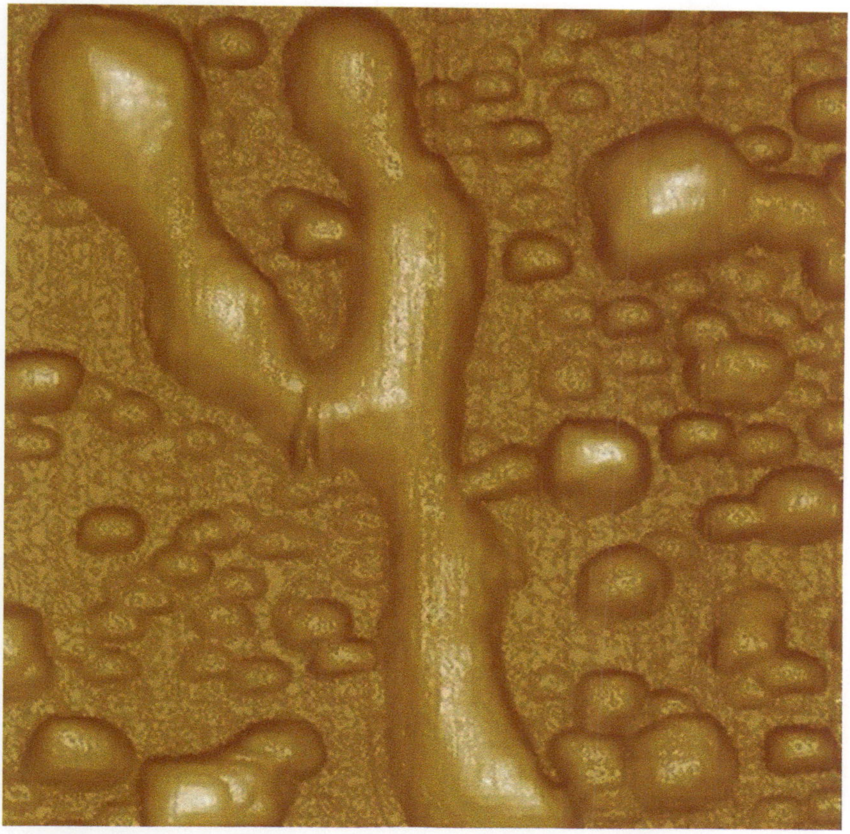

FIGURE 3.23 The Y-shaped *Mycobacterium tuberculosis* in totally drug-resistant populations recorded by the height mode of atomic force microscopy. The mechanisms involving branching in the Y-shaped tuberculosis bacilli may be a simple type of vegetative fission rather than true mycelium formation. However, we recently reported the mycelium type of division in totally drug-resistant tuberculosis bacilli.

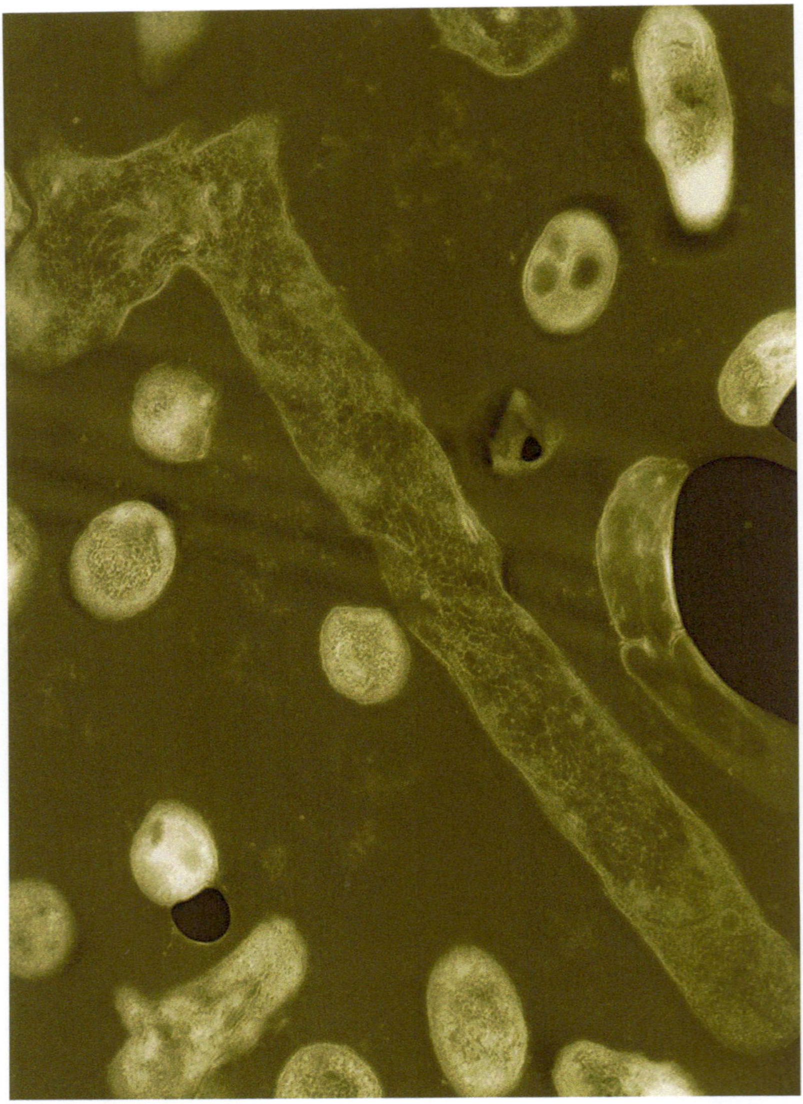

FIGURE 3.24 Elongated L-shaped *Mycobacterium tuberculosis* in the exponential phase. The dividing *M. tuberculosis* cells display a greater morphological diversity than is generally regarded. They may also appear as an elongated L-form shape.

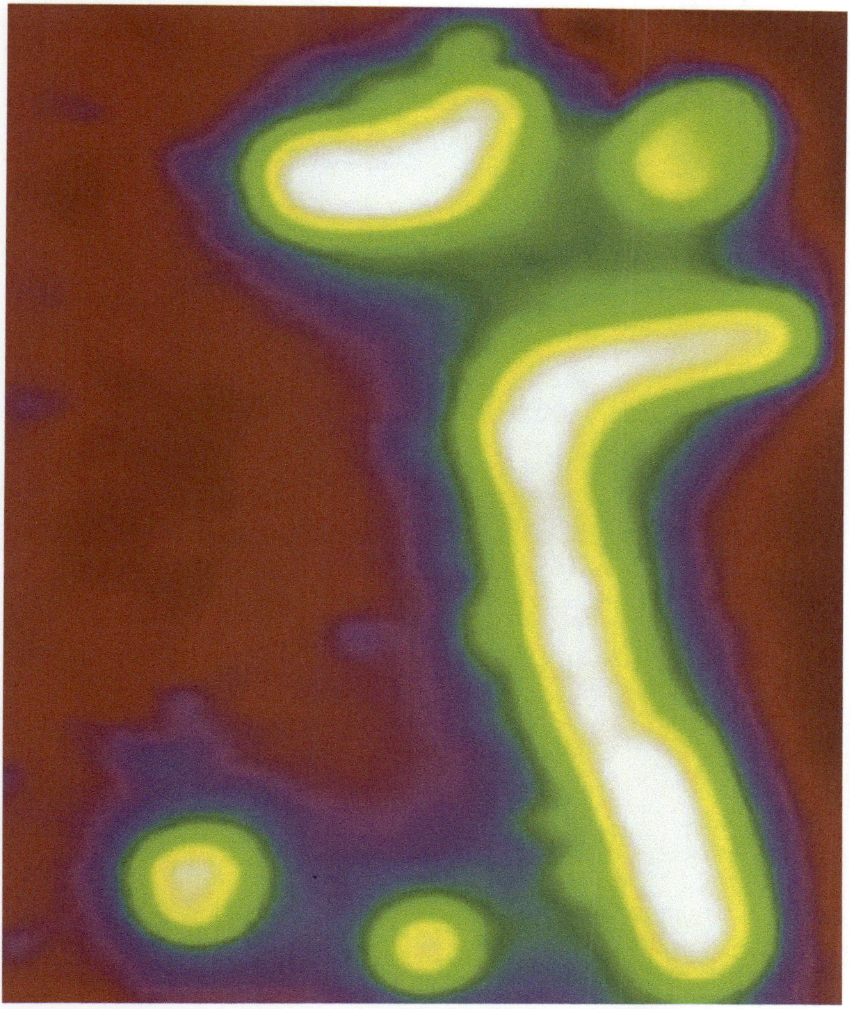

FIGURE 3.25 Variable shapes of tubercle bacilli in exponential phase recorded by atomic force microscopy. They can be of L-form, round, and rod shapes.

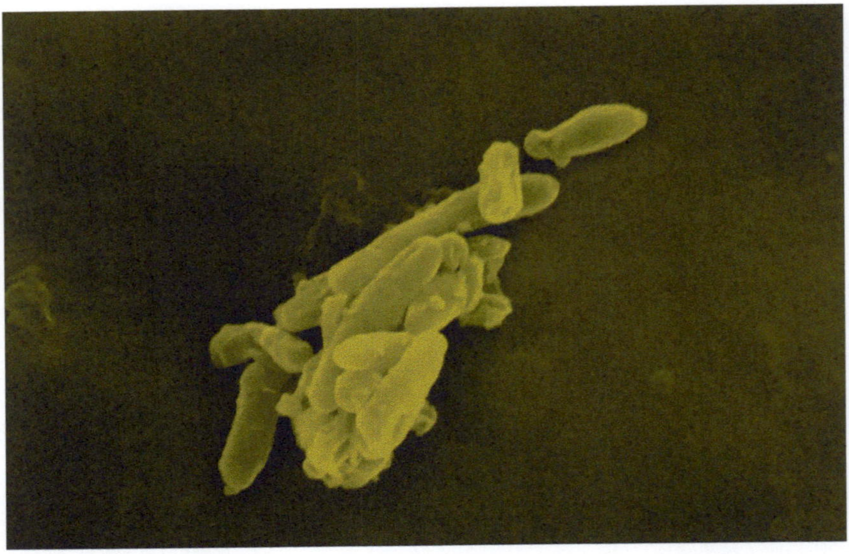

FIGURE 3.26 Scanning electron microscopy shows the variable size of *Mycobacterium tuberculosis* at the exponential phase of growth.

FIGURE 3.27 The Y-shaped *Mycobacterium tuberculosis* recorded by atomic force microscopy.

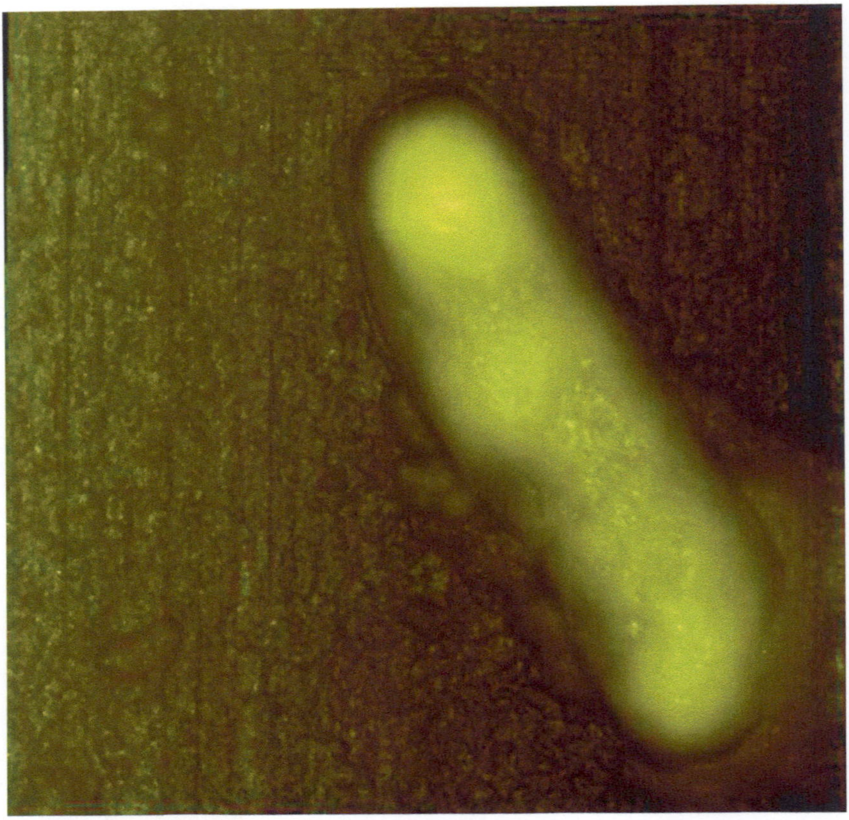

FIGURE 3.28 The classical rod shape of *Mycobacterium tuberculosis* recorded by atomic force microscopy.

FIGURE 3.29 The oval-shaped *Mycobacterium tuberculosis* recorded by atomic force microscopy.

Chapter 4

Cell Division in *Mycobacterium tuberculosis*

The Actinomycetes make up a morphologically diverse family including filamentous, coccoid, rod-shaped, and fruiting-body–producing bacteria, many of which display unique and complex life cycles (Embley and Stackebrandt, 1994; Goodfellow et al., 2012). In general, the rod-shaped bacteria grow along their longitudinal axis up to a critical size, after which they divide into two identical progeny by binary fission (Harry et al., 1999; Trevors, 2004; Goehring and Beckwith, 2005; Donovan and Bramkamp, 2014). This reproduction follows an ordered, programmed sequence of events in which cells increase in mass, replicate genetic material, and segregate newly formed chromosomes before dividing the cytoplasmic and genetic material into two daughter cells. Therefore in many rod-shaped bacteria the cell division is divided into two stages of process: (1) elongation of mother cells and (2) division of elongated mother cells into two equal cells. In this symmetrical manner of division, the daughter cell has the same size. However, mycobacteria do not always adhere to the "one size fits all" rule and can divide in an asymmetric manner that produces daughter cells of unequal sizes (Typas et al., 2010; Kieser and Rubin, 2014; Uhía et al., 2015).

The earliest observable events in cell division in *Mycobacterium* species started with the tremendous work of Brieger and Fell (1945) on the *Mycobacterium avium* life cycle. They described the growth of *M. avium* on chicken embryo extract as follows: "The *bacillus* elongates during the first 24h to form a filament, several times its original length; this divides repeatedly to form a bunch of 20–30 filaments which on the second or third day break down into short rods, which continue to multiply slowly by ordinary fission." If the bacilli are subcultivated in fresh embryo extract, then the cycle is repeated but not otherwise. This pattern was also observed on a grid using phase-contrast microscopy (Brieger and Glauert, 1952). Later, McCarthy (1971, 1974, 1976, 1978) proposed a complex relation among bacterial morphology, growth conditions, and virulence for animals. Technical advances in electron microscopy revealed the importance of the intracytoplasmic membrane during cell division in *Mycobacterium leprea*, *Mycobacterium lepramurium*, and other *Mycobacterium* species (Wheate, 1951; Yamamoto et al., 1958; Imaeda and Convit, 1962; Imaeda and Ogura, 1963; Edwards, 1970).

Further studies documented the sequential changes during bacillary division in *M. leprea* (Edwards, 1970). The observed changes in *Mycobacterium* species during cell divisions included (1) the development of a slight concavity in the plasma membrane, (2) the formation of two new electron-dense layers between the concavity and the original cell wall, and (3) the invagination or annular growth of the plasma membrane and cell wall until division is complete (Werner, 1951; Brieger and Glauert, 1952; Edwards, 1970). Kölbel (1970) proposed the theory of conjugation for mycobacteria. However, alternative interpretations of the images, either as accidentally adjacent bacteria or as branched cells, have not been ruled out. Csillage (1962, 1964, 1970) described *Mycobacterium tuberculosis* as dimorphic organisms and submitted an argument for the unlikelihood of contamination by isolating mycococci from cultures of *Mycobacterium phlei*. From published pictures of other species it seems that the process of cell division in mycobacteria resembles that of most gram-positive bacteria. A centripetal invagination of the plasma membrane is followed by formation of a cross-wall that splits to form the poles of the two new cells (Burdett and Higgins, 1978; Rastogi and David, 1981; Takade et al., 1983). Now with advances in cell biology techniques, visualization and understanding of the fundamental processes in cell division become possible (Dahl, 2004; Stewart et al., 2005; Thanky et al., 2007; Aldridge et al., 2012; Singh et al., 2013; Kieser and Rubin, 2014; Donovan and Bramkamp, 2014; Cameron et al., 2015). The main steps in the symmetrical type of cell division in *M. tuberculosis* include (1) DNA replication and segregation, (2) FtsZ ring assembly and maturation, (3) septal invagination with constriction of the envelope layer, and (4) septum closure and splitting of daughter cells. FtsZ is a small prokaryotic cytoskeletal protein analog of eukaryotic tubulin, which polymerizes as a ring preselected site acting as a scaffold for assembly of the division machinery, also known as the divisome (Bi and Lutkenhaus, 1991; Harry, 2001; Dajkovic and Lutkenhaus, 2006; Barák and Wilkinson, 2007). The Z-ring recruits all of the downstream components of the divisome and participates in the complete division process (Dziadek et al., 2003; Chen and Beckwith, 2001; Hett and Rubin, 2008) (Figs. 4.1–4.25).

In *Mycobacterium* spp. the components of the division machinery and their assembly at the Z-ring are similar to that of other rod-shaped bacteria, but specific features are present that distinguish them from other organisms. These features include (1) distribution of Z-ring positions from the middle of the cell to near the poles in dividing cells, thus the division sites are more dispersed than those in *Escherichia coli* or *Bacillus subtilis*; (2) absence of nucleoid occlusion (ie, septum placement independent of nucleoid position); (3) active translocation of chromosomal DNA during segregation; (4) unequal growth at the two poles; and (5) different rates of growth for unequal siblings (Chauhan et al., 2006; Thanky et al., 2007; Kang et al., 2008; Singh et al., 2012; Vijay et al., 2014). Joyce et al. (2012) proposed that asymmetry in division is a consequence of unequal polar growth after division site placement. Singh et al. (2013) attributed this to the unequal growth at the two poles after Z-ring formation. Recently, unusual cell cycles were correlated with resistant patterns of *M. tuberculosis* (Farnia et al., 2010). More than 40% of the total population of cells, in highly drug-resistant strains of tuberculosis, showed a multibranching, budding, or asymmetrical type of division. Aldridge et al. (2012) and Wakamoto et al. (2013) have modeled how asymmetric growth and division influence cell fate during antibiotic therapy in vitro studies. Furthermore, it is suggested that there is a correlation between an unusual type of cell division in *M. tuberculosis* and its ability to persist in a quiescent state (Barry et al., 2009; Velayati et al., 2012a,c) (Figs. 4.26–4.54).

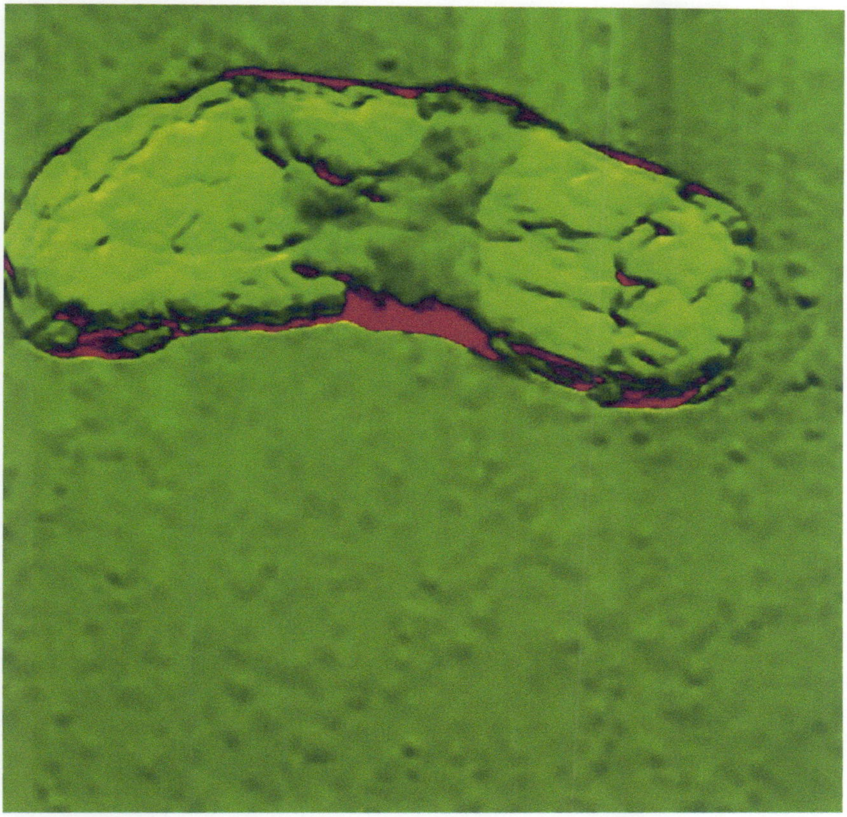

FIGURE 4.1 *Mycobacterium tuberculosis* in exponential phase of growth. The extensive heterogeneity is observed in the exponential phase of growth. The average size of *M. tuberculosis* is from 1 to 3 μm.

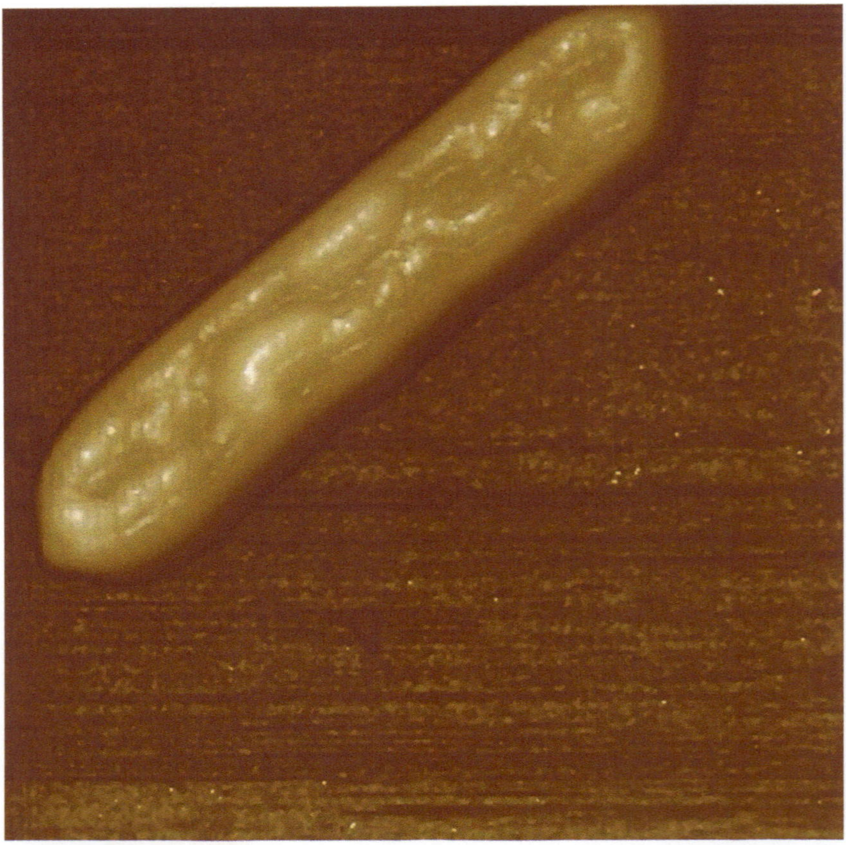

FIGURE 4.2 *Mycobacterium tuberculosis* undergoing the division cycle. In general, the cell cycle is the sum of biosynthetic patterns of cytoplasm, the genome, and the cell surface.

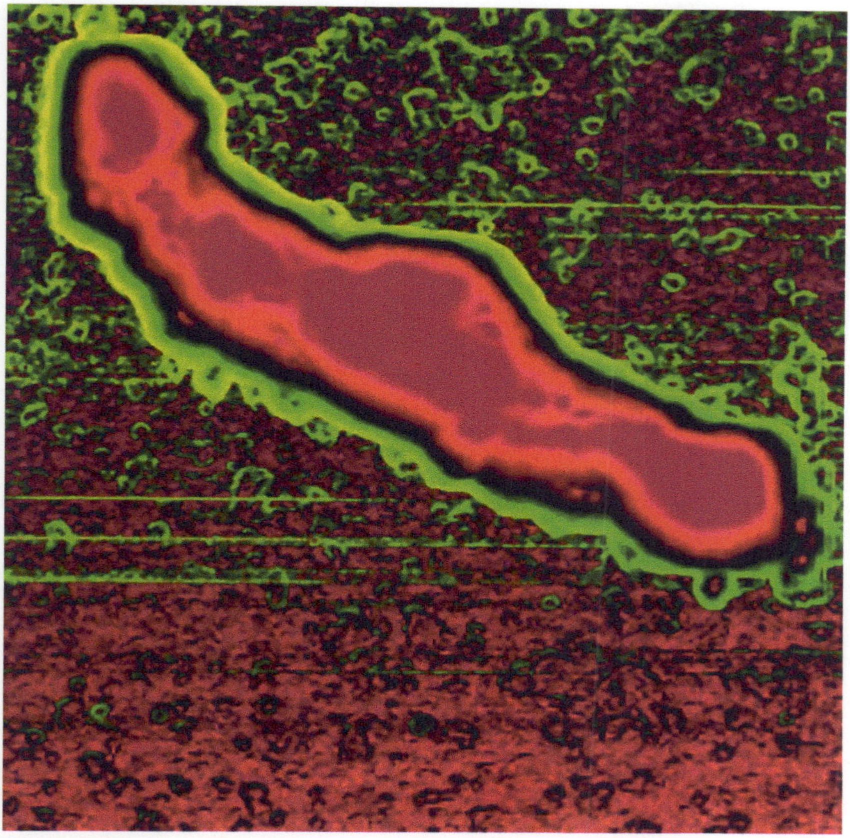

FIGURE 4.3 *Mycobacterium tuberculosis* in exponential cytoplasm synthesis during the division cycle, recorded by atomic force microscopy. Each unit of cytoplasm produces a small amount of new cytoplasm over a small interval of time. In this process, no particular molecule of the cytoplasm is made differently from any other molecule of the cytoplasm. The rate of cytoplasm synthesis and the pattern of accumulation of cytoplasm are exponential during the cell division cycle.

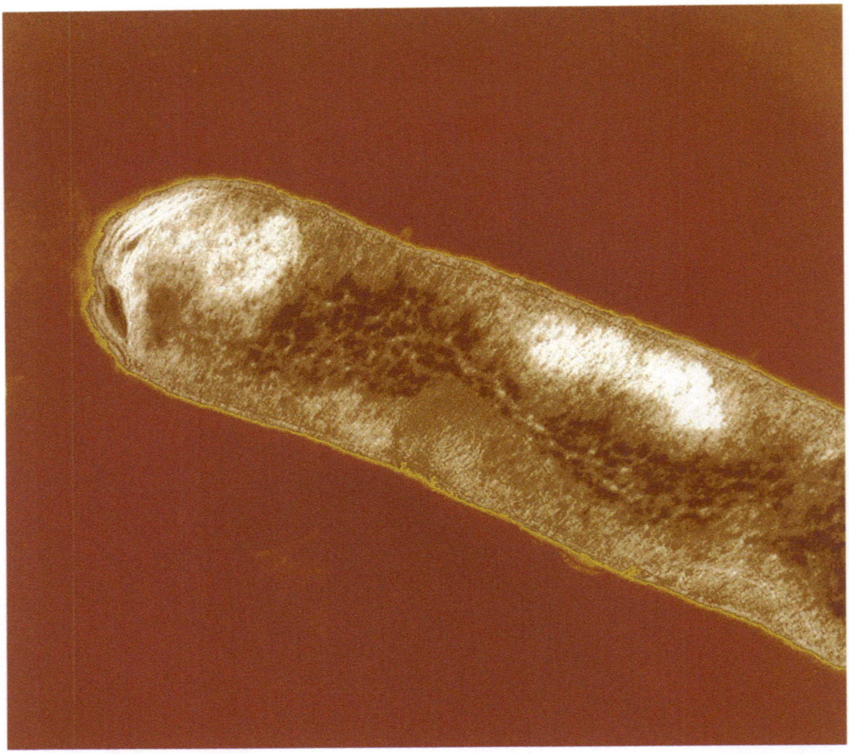

FIGURE 4.4 Longitudinal section of *Mycobacterium tuberculosis* in exponential cytoplasm synthesis under transmission electron microscopy. Before septum formation the biosynthesis of cytoplasm (which is the entire accumulation of proteins, RNA molecules, ribosomes, small molecules, water, and ions) and one-dimensional linear DNA structure have occurred.

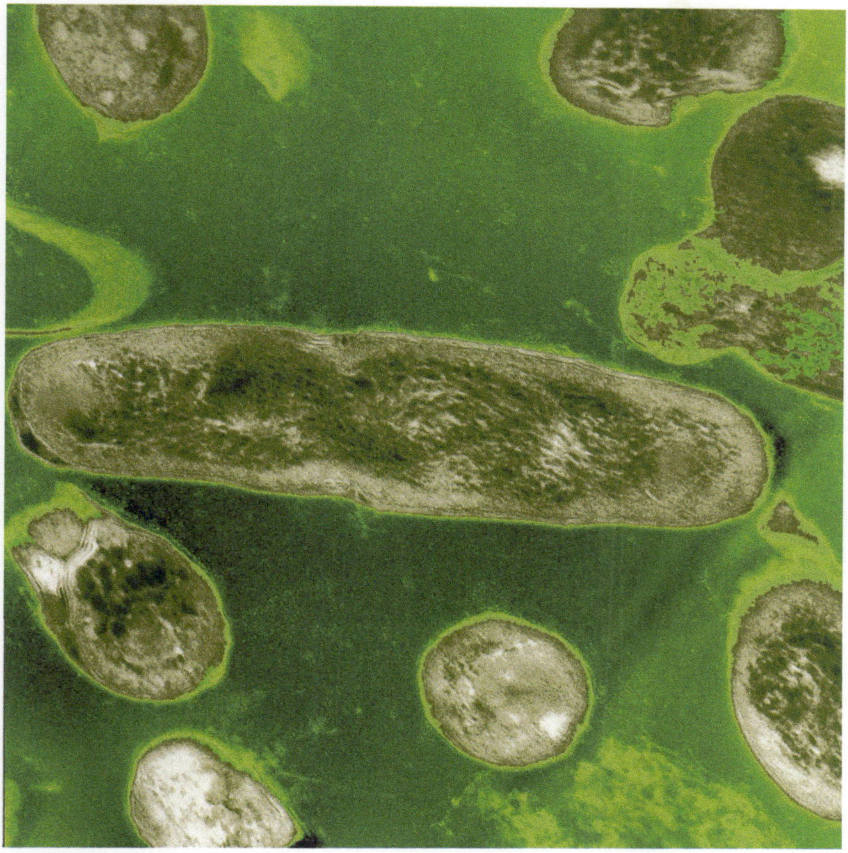

FIGURE 4.5 *Mycobacterium tuberculosis* starting the cell division cycle under transmission electron microscopy.

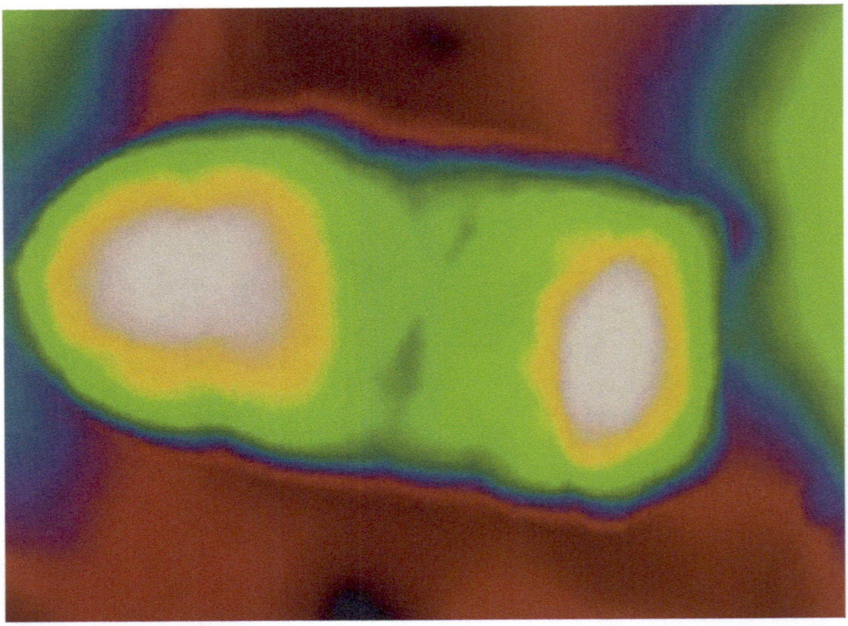

FIGURE 4.6 Bipolar growth in *Mycobacterium tuberculosis*. Polar growth occurs within the context of the cell cycle. The cytoplasmic separation began after clearance of chromosome and cytoplasmic content from the midpoint of elongated cells.

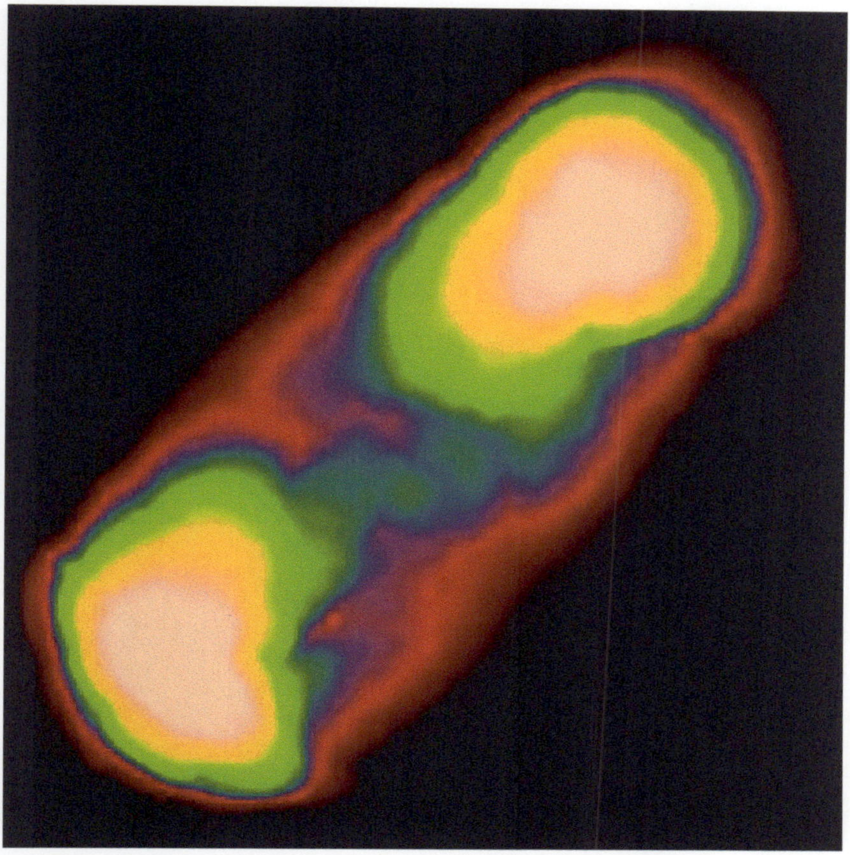

FIGURE 4.7 *Mycobacterium tuberculosis* divides into two identical progeny by binary fission.

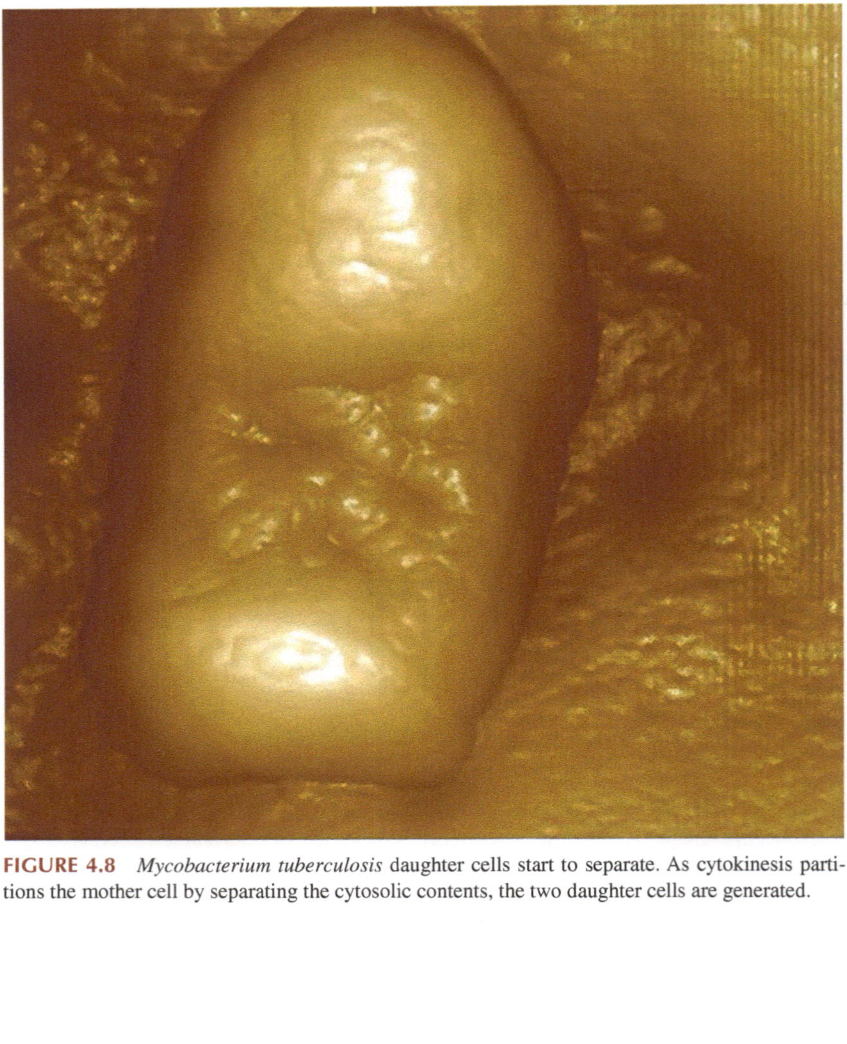

FIGURE 4.8 *Mycobacterium tuberculosis* daughter cells start to separate. As cytokinesis partitions the mother cell by separating the cytosolic contents, the two daughter cells are generated.

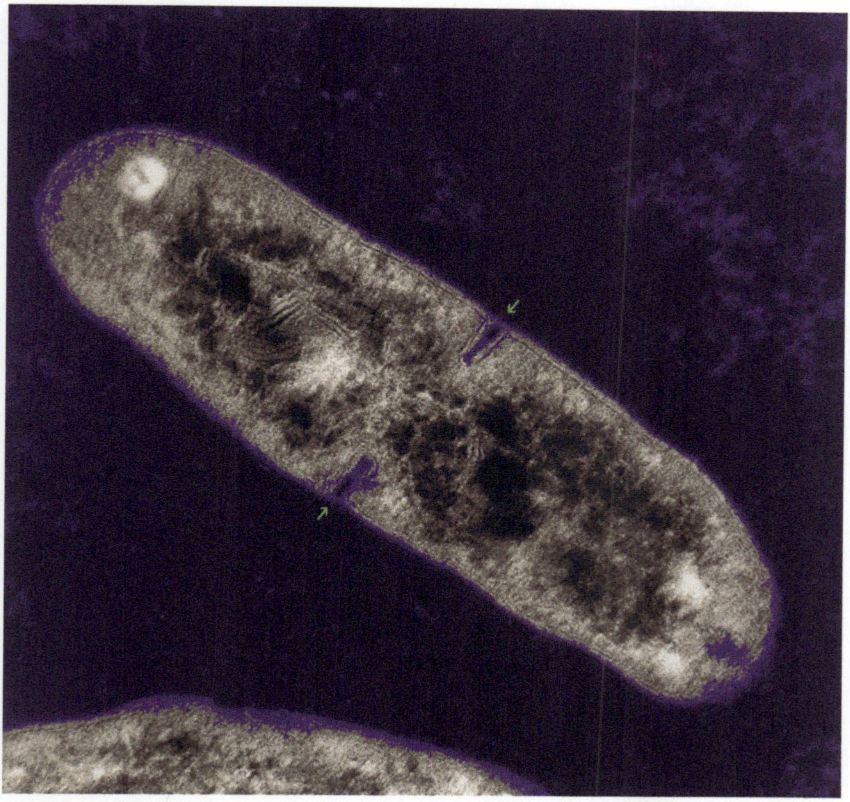

FIGURE 4.9 Longitudinal section of *Mycobacterium tuberculosis* before daughter cell separation. The invagination process or septum (*green arrows*), the formation that forms two poles in the middle of the cell, involves a complex process by which the plasma membrane and inner cell wall grow inward but the outer cell wall remains intact.

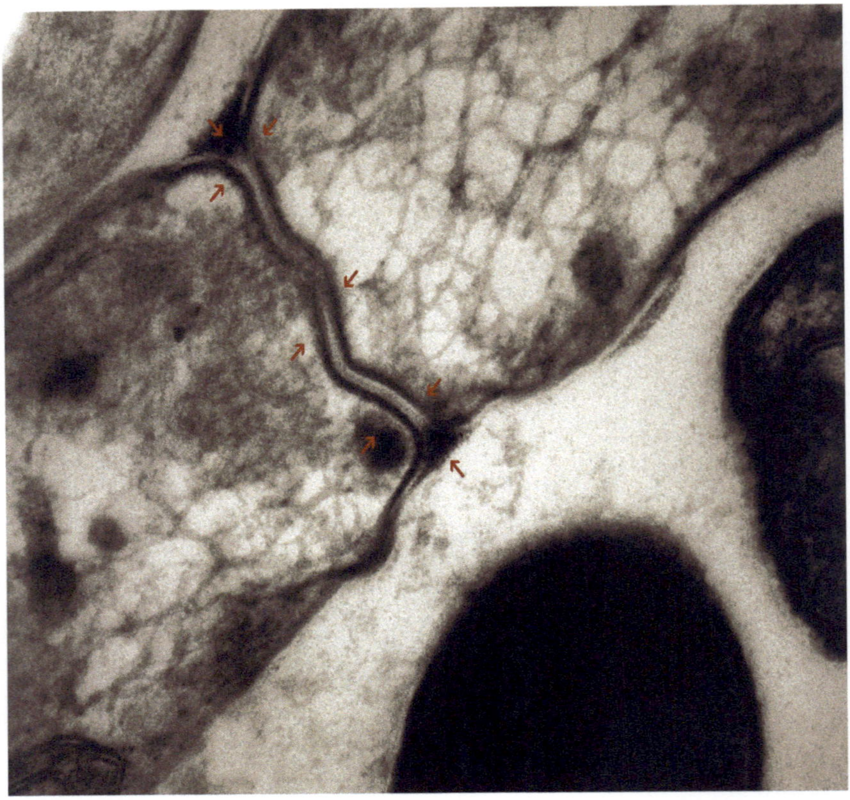

FIGURE 4.10 Cross-section of heat-killed *Mycobacterium tuberculosis* at the exponential growth phase, which has competed septum formation. The septum is composed of four components: a plasma membrane and an inner cell wall component on either side of a transparent zone, which had an average diameter of 27 nm (*red arrows*). The width of the septum is approximately 21 nm at each end and 30 nm at the center.

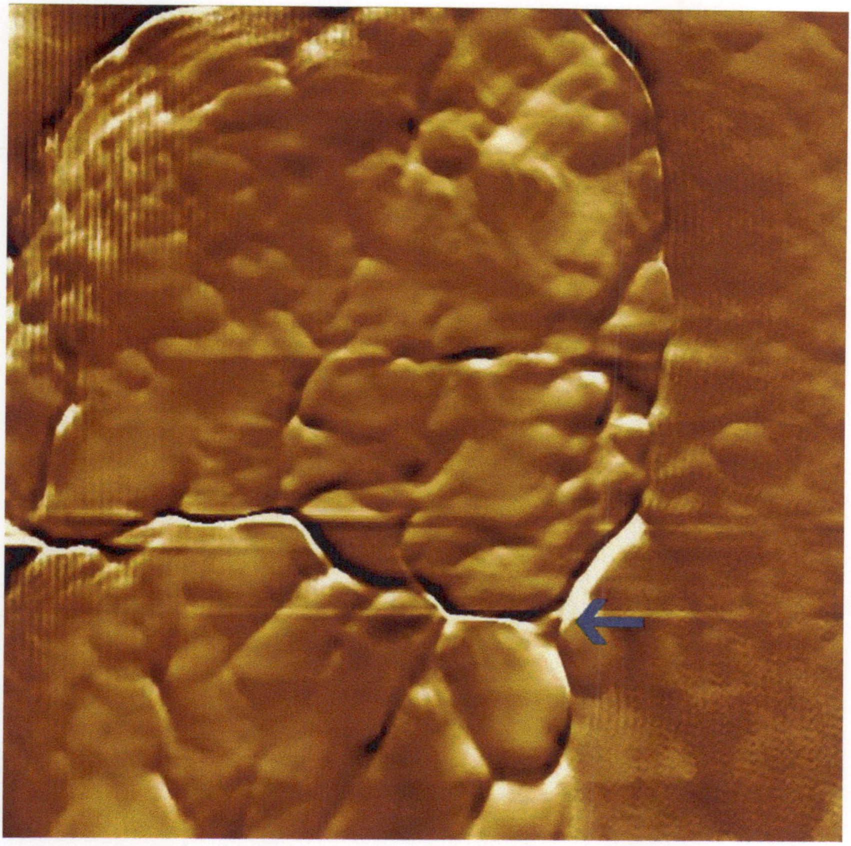

FIGURE 4.11 The separation of septal wall material (*blue arrow*) in *Mycobacterium tuberculosis* recorded by atomic force microscopy.

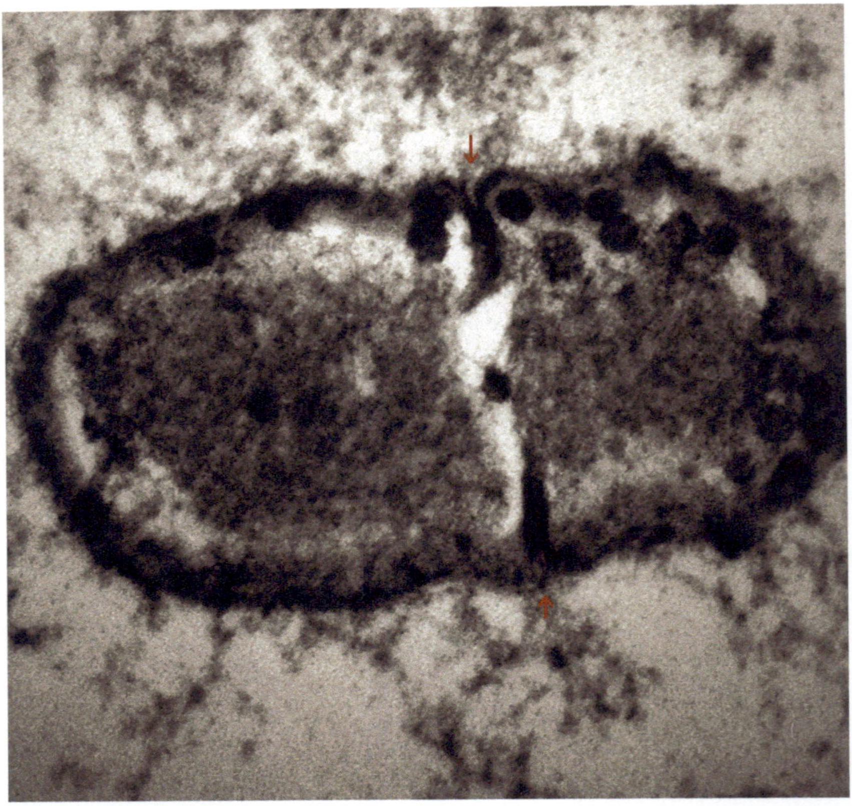

FIGURE 4.12 The separation of septal wall material (*red arrows*) in *Mycobacterium tuberculosis* under transmission electron microscopy.

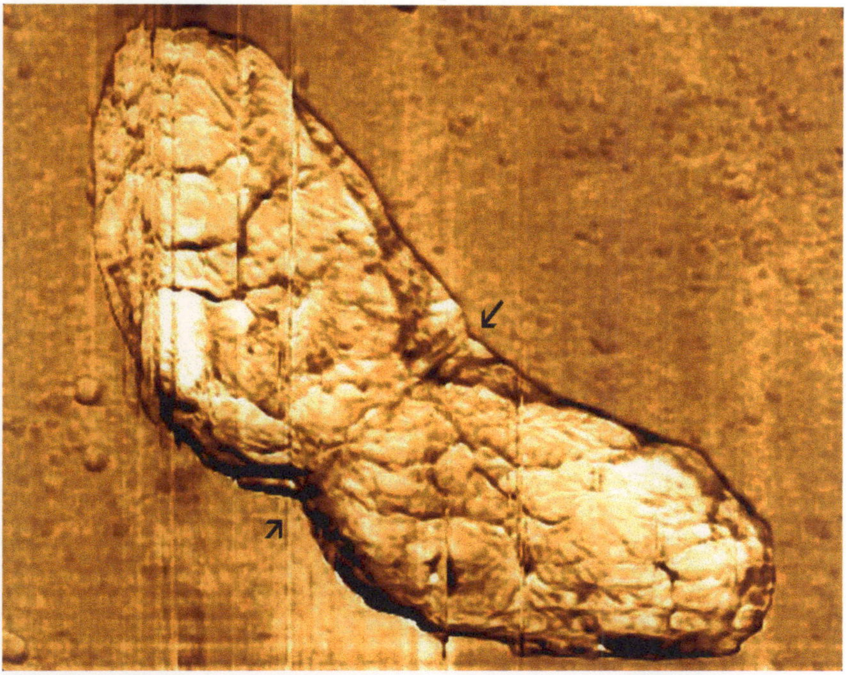

FIGURE 4.13 The septum at mid-cell of *Mycobacterium tuberculosis* (*black arrows*). In the symmetrical type of cell division, the septum formed at mid-cell, which results in two daughter cells of the same size.

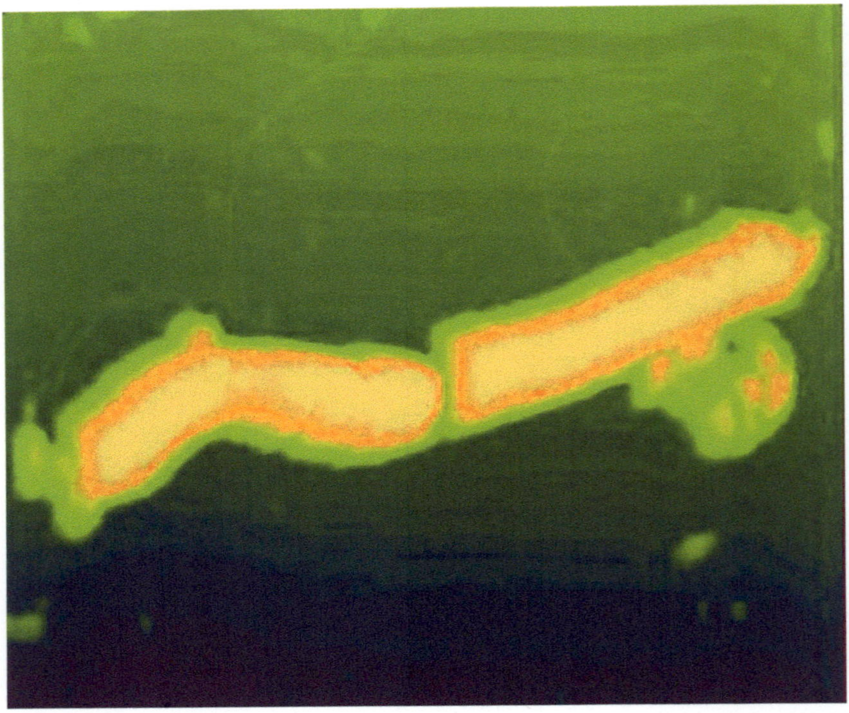

FIGURE 4.14 Symmetrical type of cell division in *Mycobacterium tuberculosis.* The precise placement of the division site is critically important, and all bacteria, including mycobacteria, must have an elaborate division site selection system. This system is composed of a large multiprotein complex that is known as the divisome. The divisome contains cytoskeletal-like proteins, structural factors, and peptidoglycan synthases and hydrolases, which coordinate the synthesis and then the splitting of the septum during cell division.

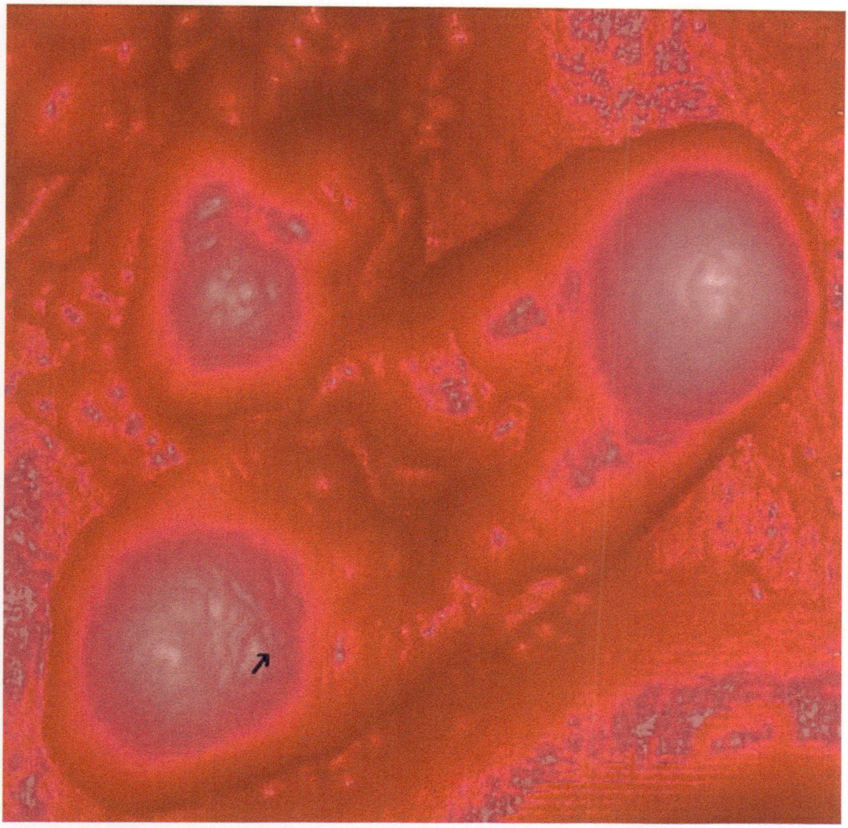

FIGURE 4.15 The banding pattern or protruding ridges on the surface of *Mycobacterium tuberculosis* (*black arrow*). One of the earliest events related to the invagination process is the localization of the FtsZ protein to the future site of pole formation. The FtsZ protein is localized to the division site in a pattern called the FtsZ ring. During the invagination process the FtsZ protein is located at the leading edge of the invagination process.

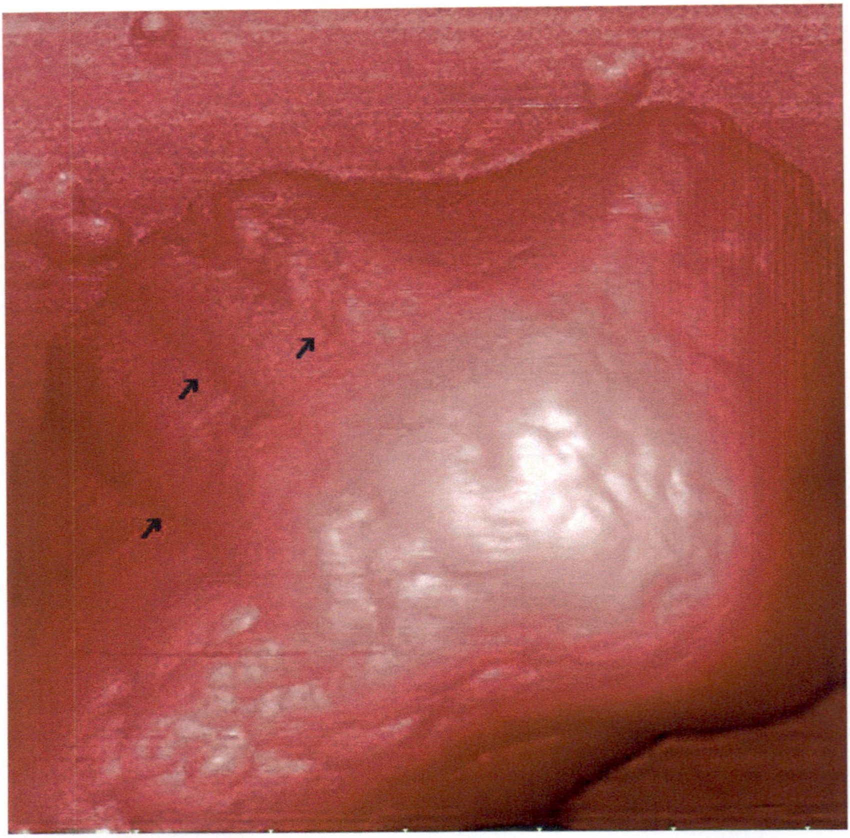

FIGURE 4.16 The external ridges on the surface of *Mycobacterium tuberculosis* (*black arrows*). The position of external ridges near the ends of cells suggests that the cells may have recently divided and these ridges are division scars. Alternatively, these circumferential bands might also be the locations for future septa formation.

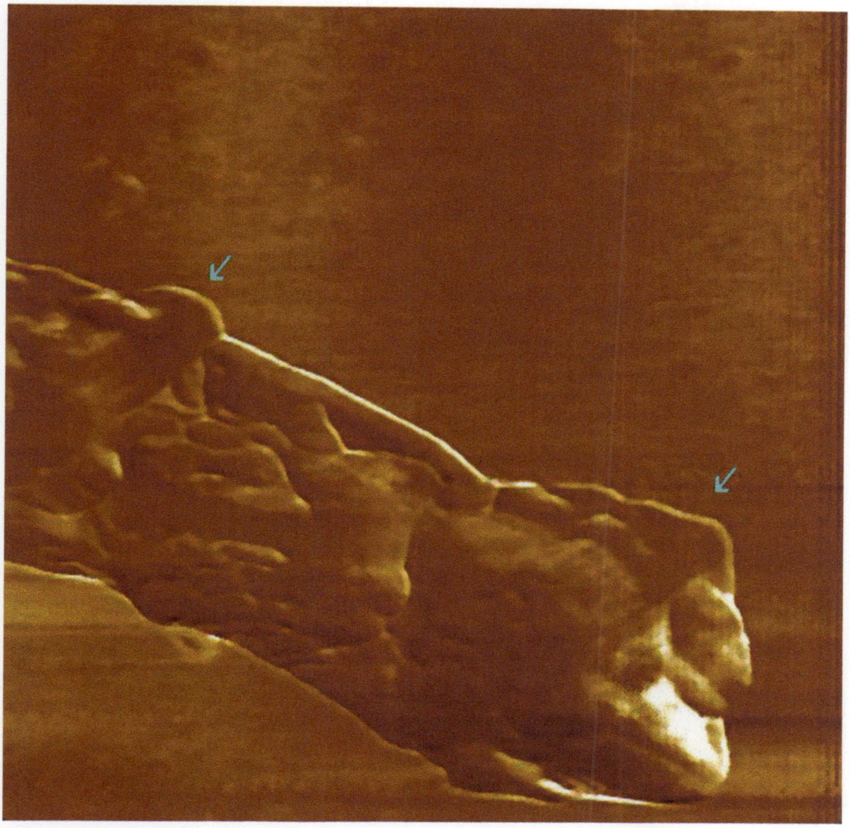

FIGURE 4.17 Multiple bands or external ridges on the surface of *Mycobacterium tuberculosis* (*arrows*). When there is multiple banding it is not clear which are the results of previous cell divisions and which bands may indicate ongoing underlying septa formation. If the ridges near the ends of the cells represent ongoing septa formation, then tubercle bacilli would be able to asymmetrically divide.

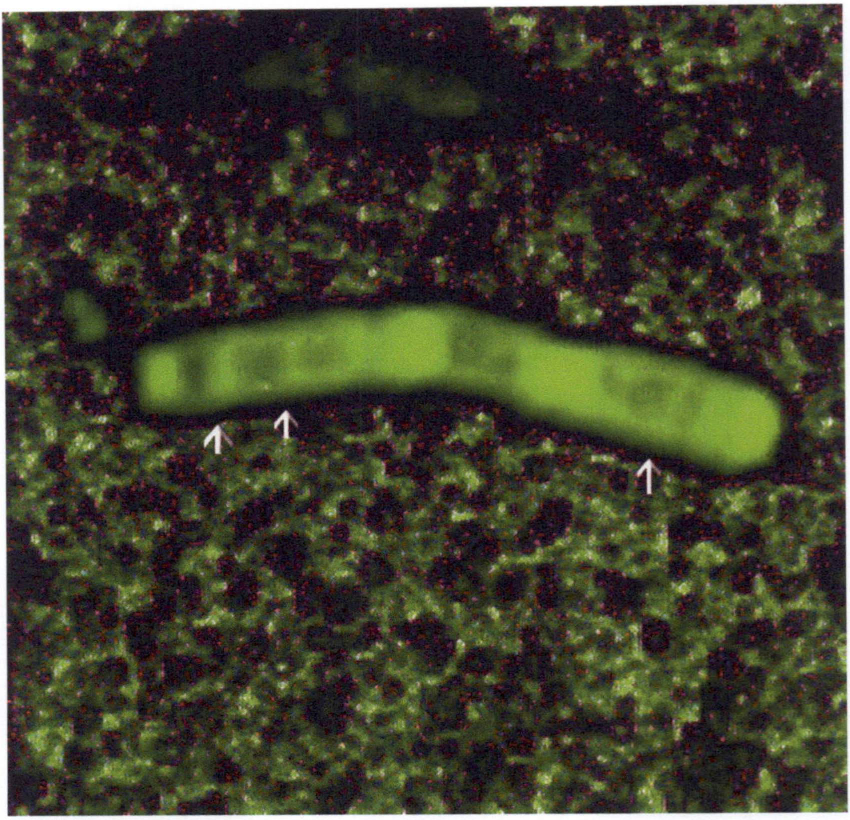

FIGURE 4.18 Elongation of *Mycobacterium tuberculosis*. It is possible for a single cell to accumulate numerous ridges as cell divisions continue (*arrows*).

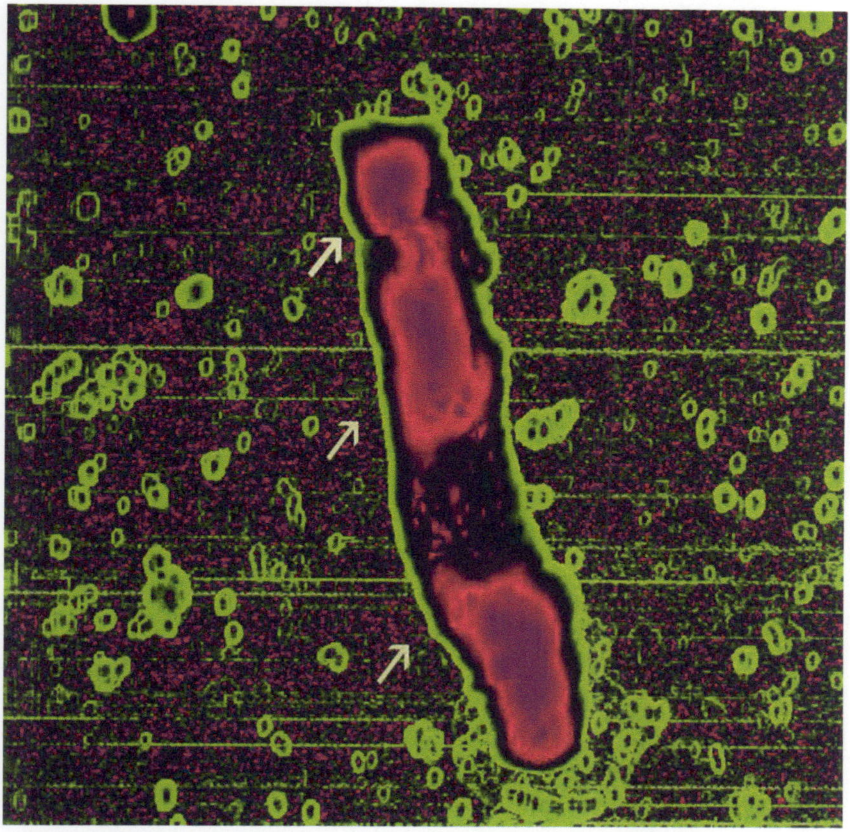

FIGURE 4.19 Multilocalization of the division site in *Mycobacterium tuberculosis* (*arrows*). The site location is strongly influenced by the existence of mini-cells. If invagination occurs at multiple sites (ie, near a pole and in the center), then a very small DNA-less mini-cell is produced.

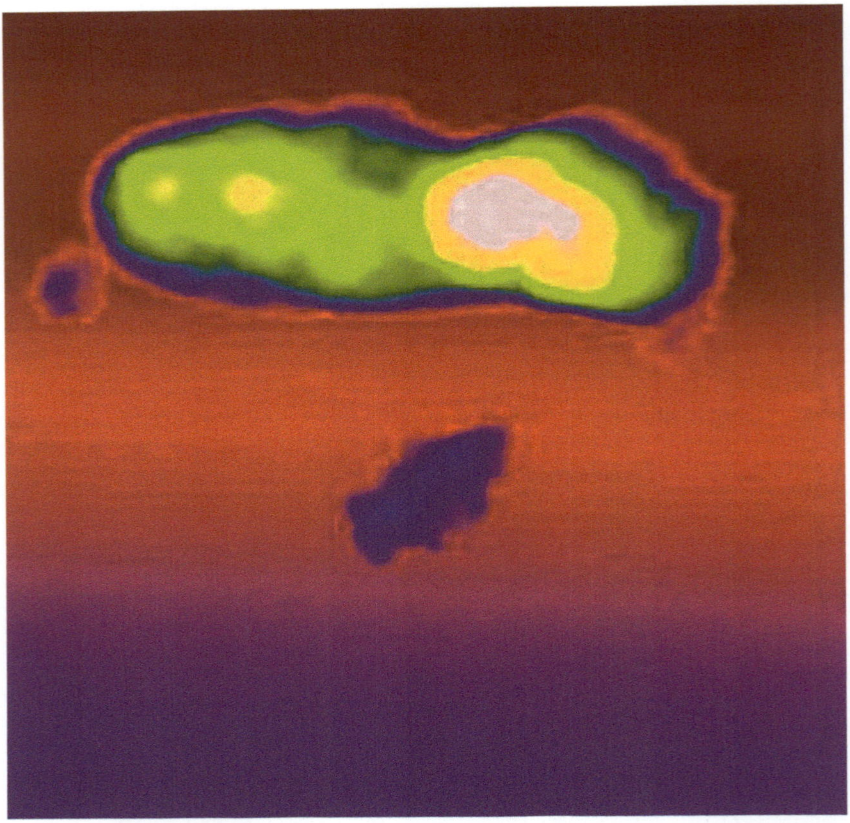

FIGURE 4.20 During balanced growth, very few DNA-less *Mycobacterium tuberculosis* daughter cells are produced. The new progeny in the right side of the bacillus has enough nuclear material (*white color*) whereas the new progeny in the left side has much less nuclear content.

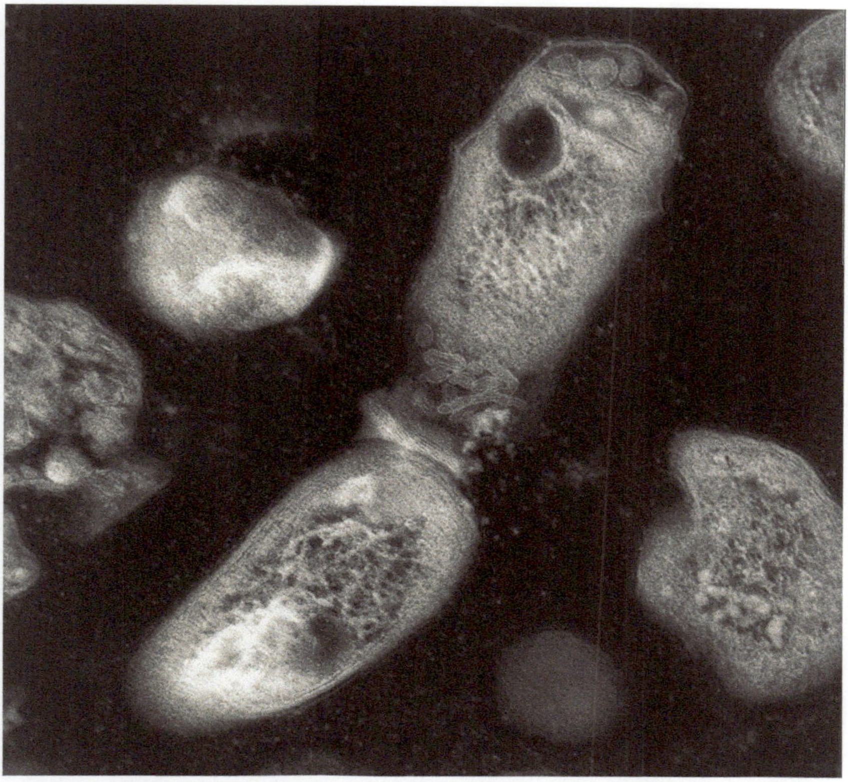

FIGURE 4.21 Segregation of the cell components at the division cycle in *Mycobacterium tuberculosis*. In a cell that divides to produce two equivalent daughter cells, the segregation problem is to ensure that both cells have the same amount of each portion of the cell material. The cytoplasm components appear to segregate randomly at division, whereas the segregation of DNA is a regulated and deterministic process.

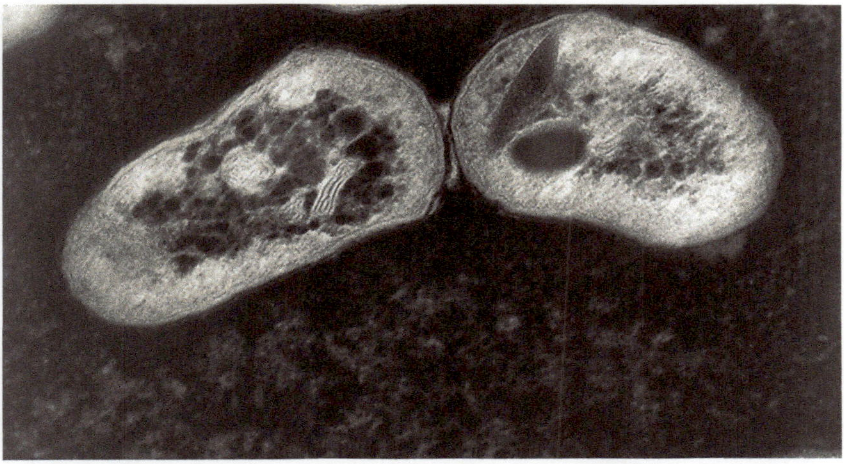

FIGURE 4.22 Separation of daughter cells at the division cycle in *Mycobacterium tuberculosis*. In general, the cell length is heterogeneous in the division cycle, but the cell width seems to be uniform. This may be due to the retention of unsalvageable peptidoglycan in the polar cap.

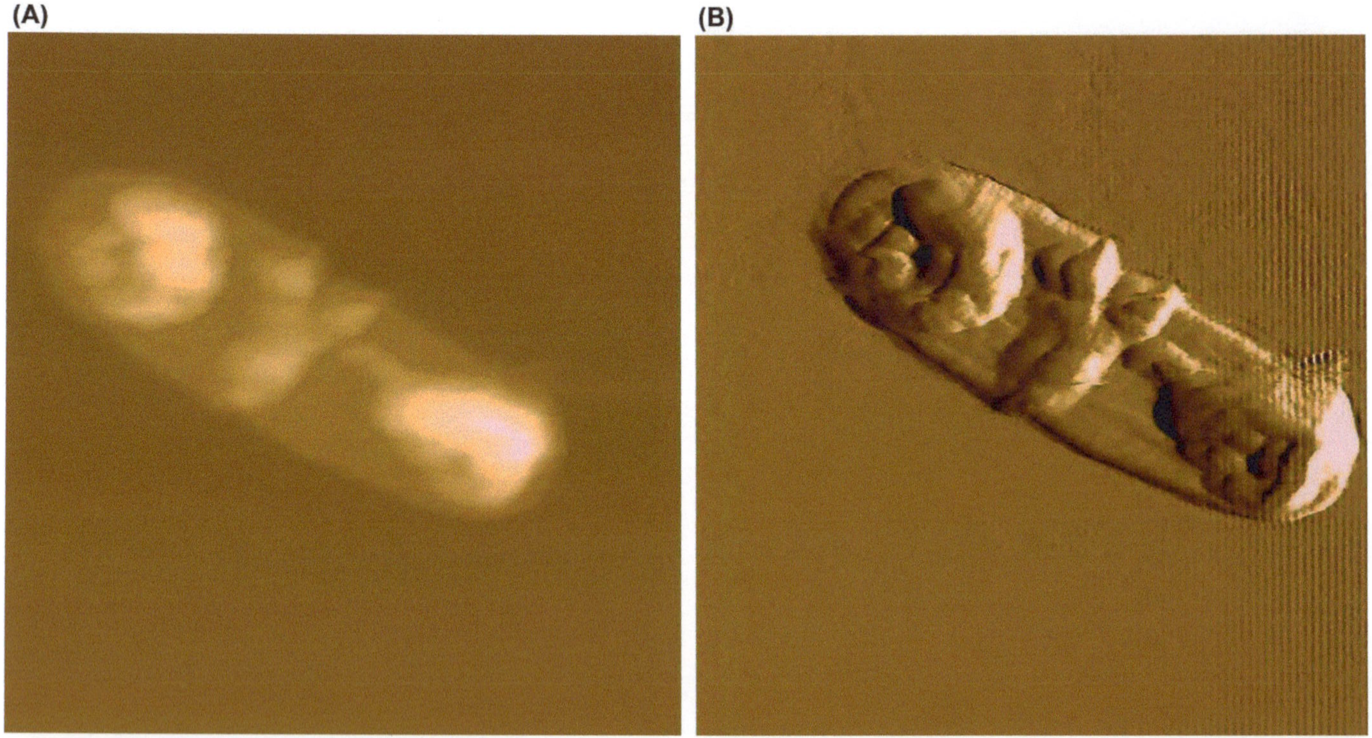

FIGURE 4.23 Septum cleaving in *Mycobacterium tuberculosis* recorded by (A) height and (B) friction mode by atomic force microscopy. After the completion of septum synthesis the septum is cleaved to physically separate daughter cells. Mycobacteria have an essential peptidoglycan hydrolase, RipA. RipA interacts with another peptidoglycan hydrolase, known as resuscitation promoting factor B (RpfB). These two peptidoglycan hydrolases synergistically cleave peptidoglycan. This synergy is counteracted by the peptidoglycan synthase PonA1. The balance between PonA1–RipA and RipA–RpfB interactions may be used as a mechanism to ensure that the synthesis of the septum precedes its hydrolysis.

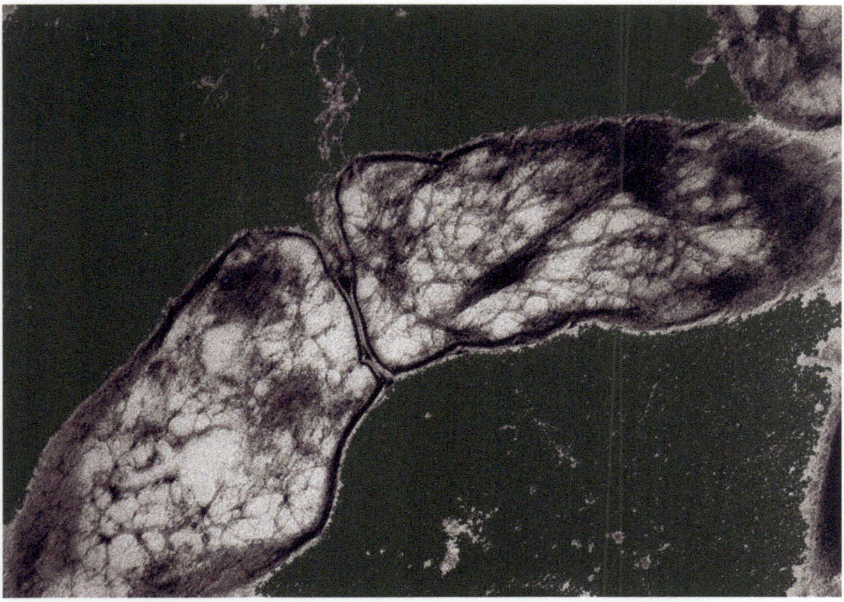

FIGURE 4.24 Septum cleaving in *Mycobacterium tuberculosis* recorded by transmission electron microscopy.

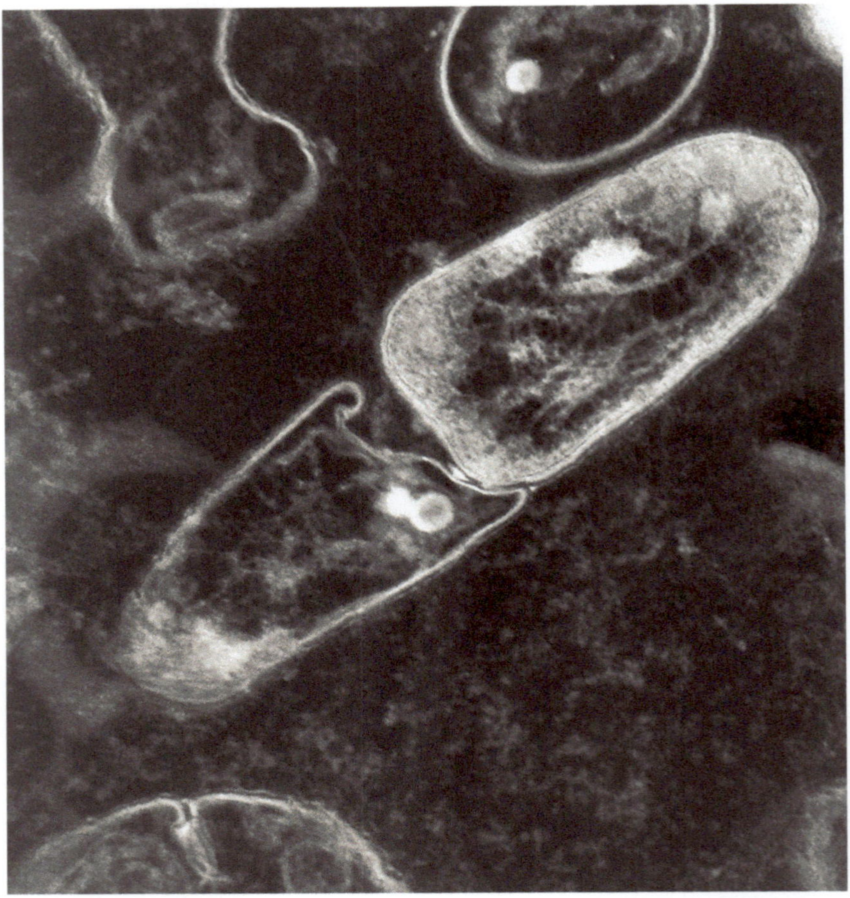

FIGURE 4.25 Daughter cells are separating at division cycle in *Mycobacterium tuberculosis*.

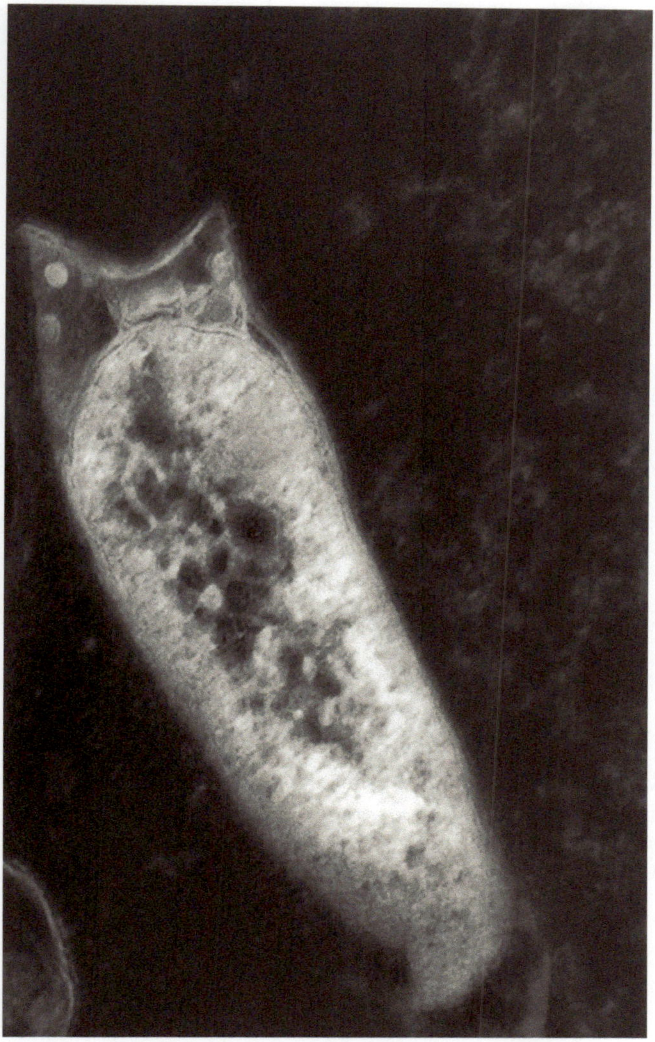

FIGURE 4.26 Zonal cell elongation in one pole in *Mycobacterium tuberculosis*. Polar growth is thought to facilitate diversification of cell shapes, growth plans, and ecological niches. Similar to many rod-shaped bacteria, *M. tuberculosis* contains two poles: an old pole (which is inherited from the mother cell) and a new pole (which is synthesized during division).

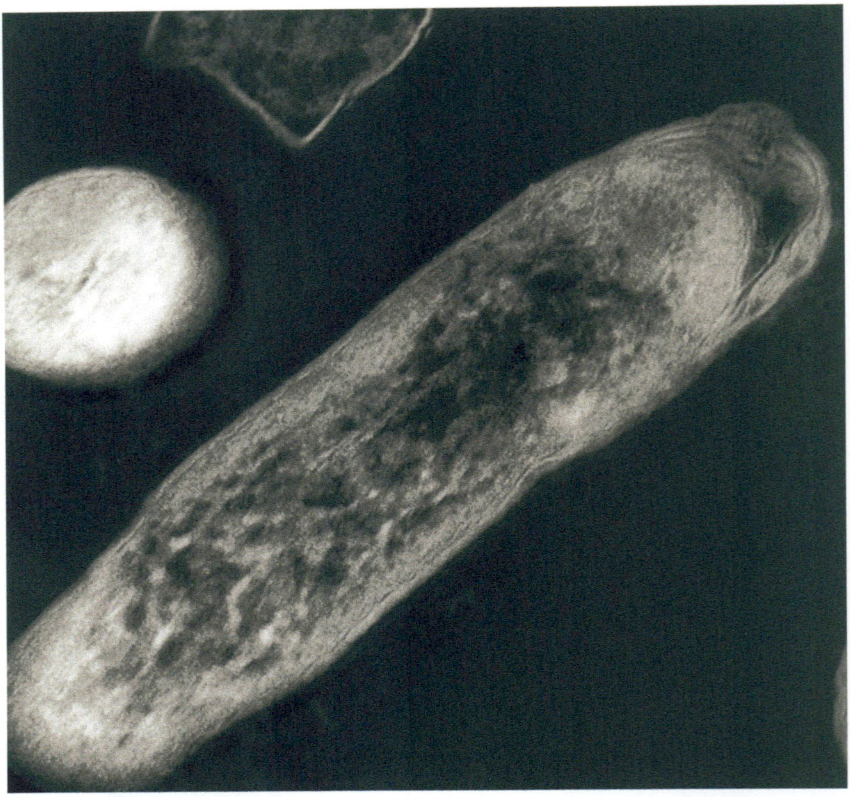

FIGURE 4.27 Starting of new pole formation in *Mycobacterium tuberculosis*. Polar growth requires the simultaneous synthesis of peptidoglycan, arabinogalactan, and mycolic acids. To ensure that elongation is directed to the cell poles, cytoplasmic proteins anchor the elongation complex at these sites.

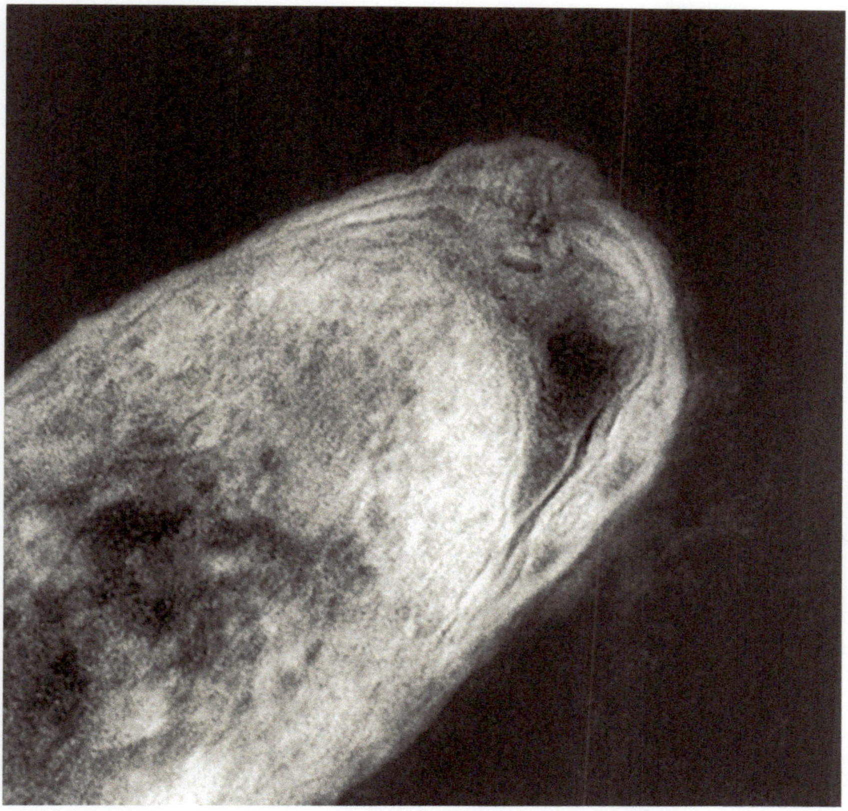

FIGURE 4.28 New pole formation in *Mycobacterium tuberculosis*.

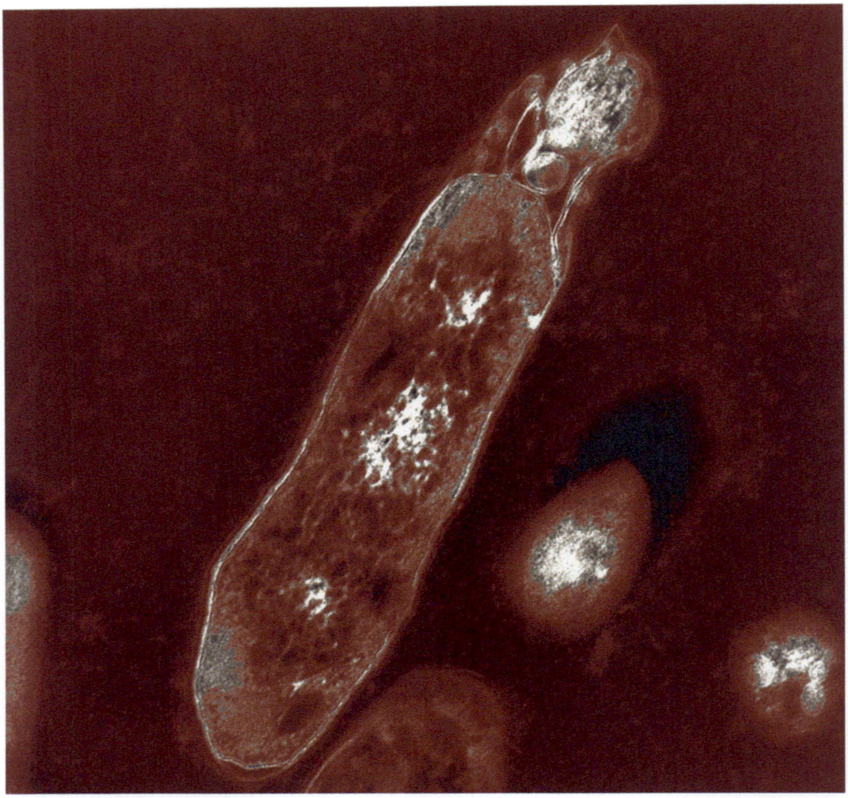

FIGURE 4.29 *Mycobacterium tuberculosis* elongate by incorporating new cell wall material into the new pole. The rate of elongation is frequently not equal at the old and new poles; therefore the resulting daughter cells may be of different size.

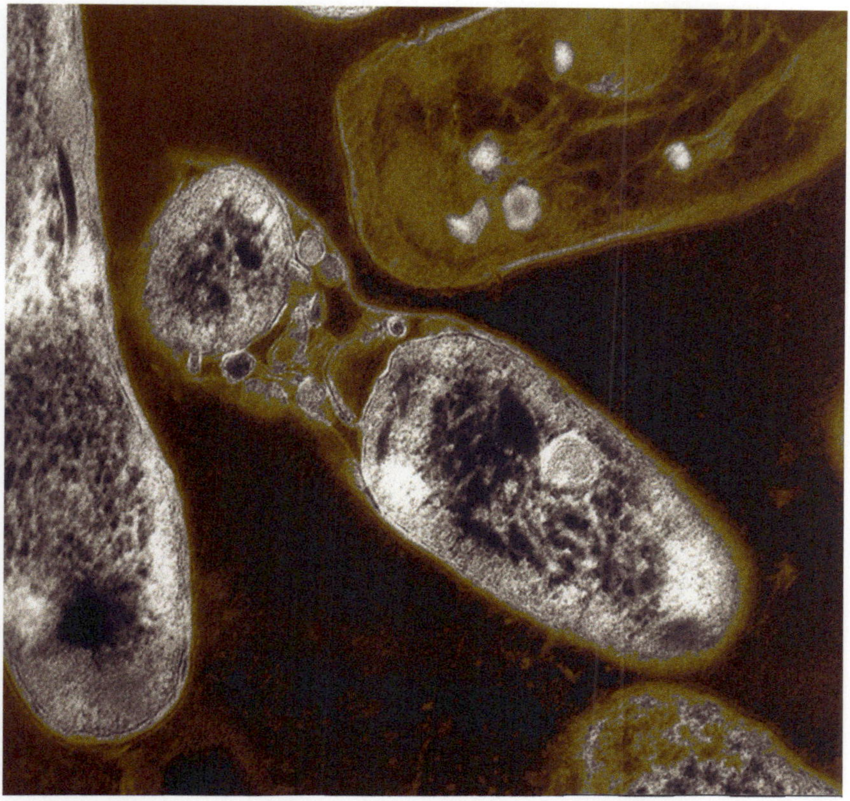

FIGURE 4.30 Asymmetric bipolar growth in *Mycobacterium tuberculosis*. This type of growth resulted in a longer old-pole cell and a shorter new-pole cell.

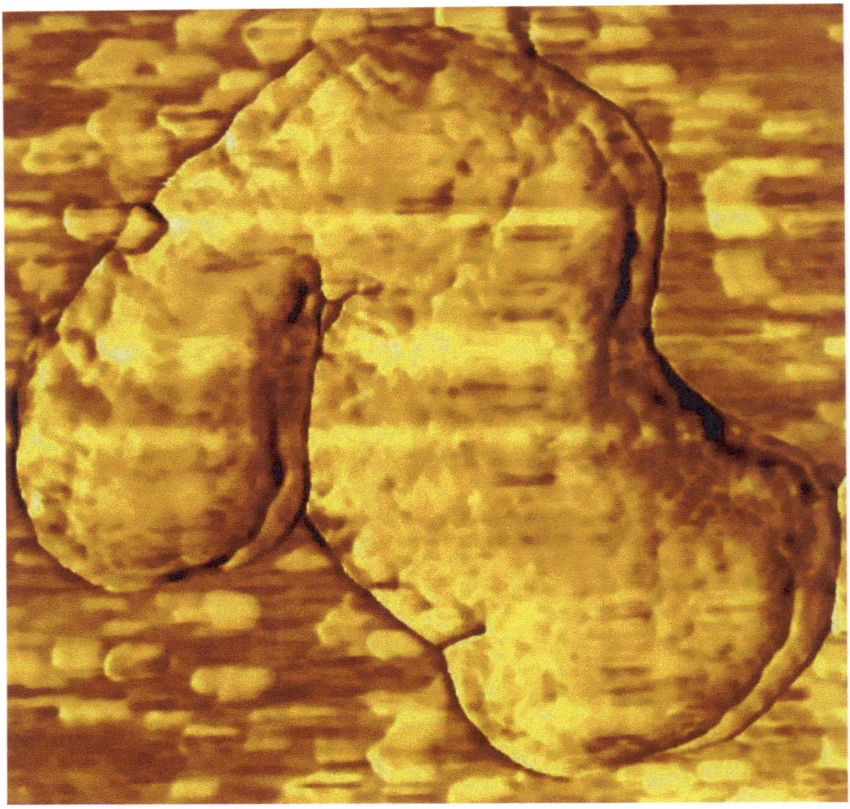

FIGURE 4.31 Asymmetric septum position in *Mycobacterium tuberculosis*. In general, any division with more than 10% deviation in septum position from the mid-cell site is considered asymmetric with an unusually deviated septum, generating a short cell and a long cell.

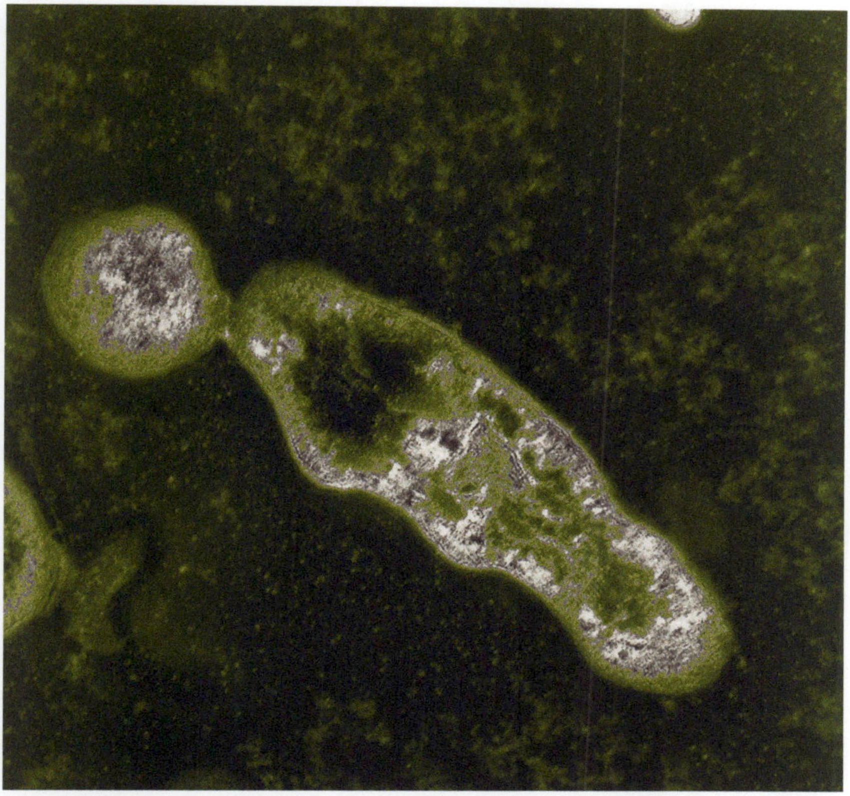

FIGURE 4.32 Transmission electron microscopy shows the round- and rod-shaped bacilli.

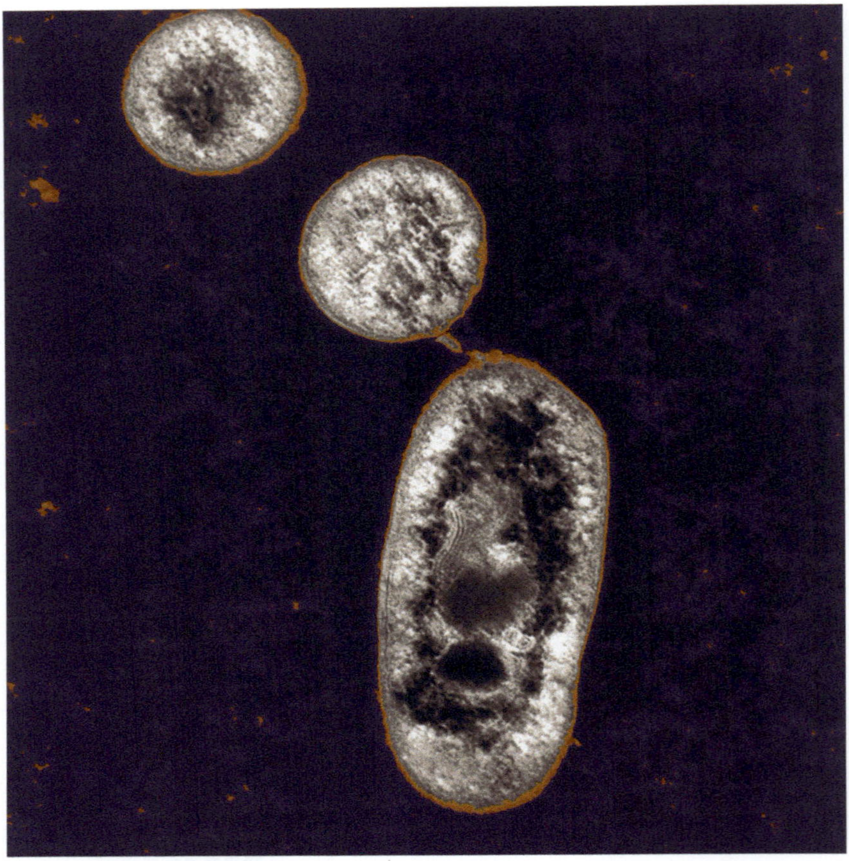

FIGURE 4.33 Transmission electron microscopy shows formation of round-shaped bacilli from rod-shaped mother bacilli.

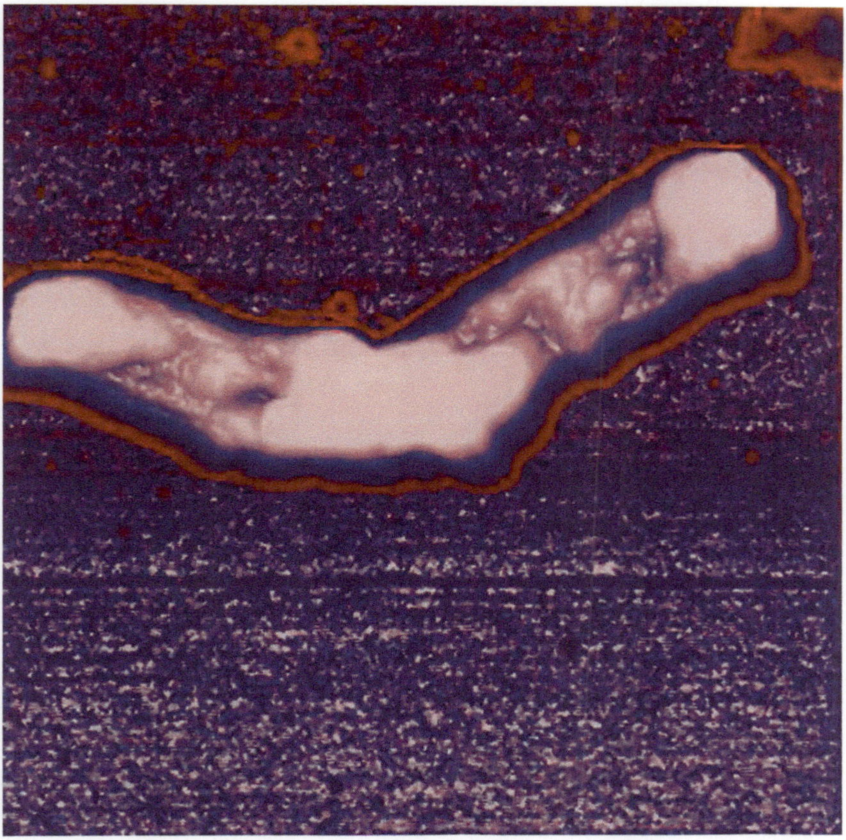

FIGURE 4.34 Snapping type of cell division in *Mycobacterium tuberculosis* recorded by atomic force microscopy. Upon completion of cell division, one or both of the two daughter cells suddenly swing around, bringing their distal ends closer together while still remaining attached by a small region at their proximal ends.

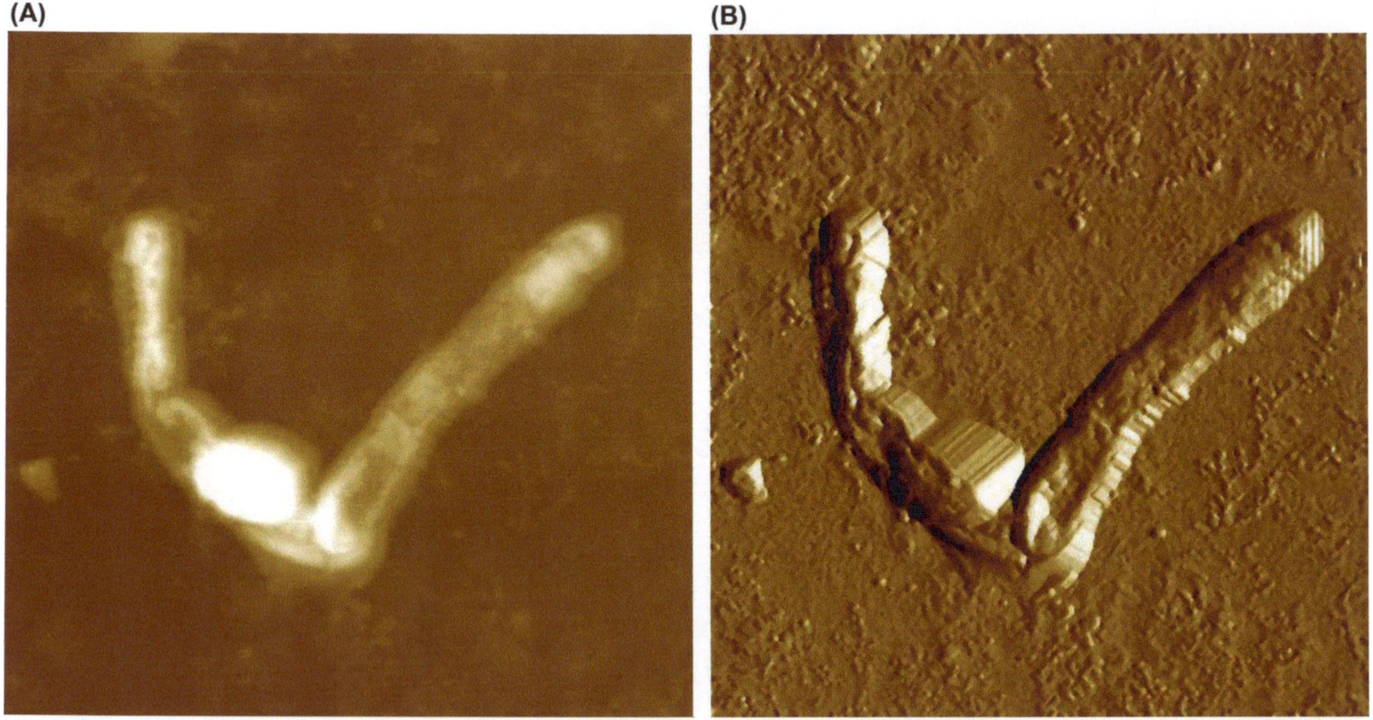

FIGURE 4.35 V-form bacilli during division cycle recorded by (A) height and (B) friction mode by atomic force microscopy. In *Mycobacterium tuberculosis*, during septum formation the plasma membrane and inner cell wall grow inward but the outer cell wall layer remains intact. Upon completion of septum formation with a cross-wall, the inner layer may continue to grow and thus exert pressure upon the outer cell wall layer. The outer layer eventually ruptures first on one side of the cell, and the two daughter cells bend in one side, where the outer layer is still intact forming a V-form.

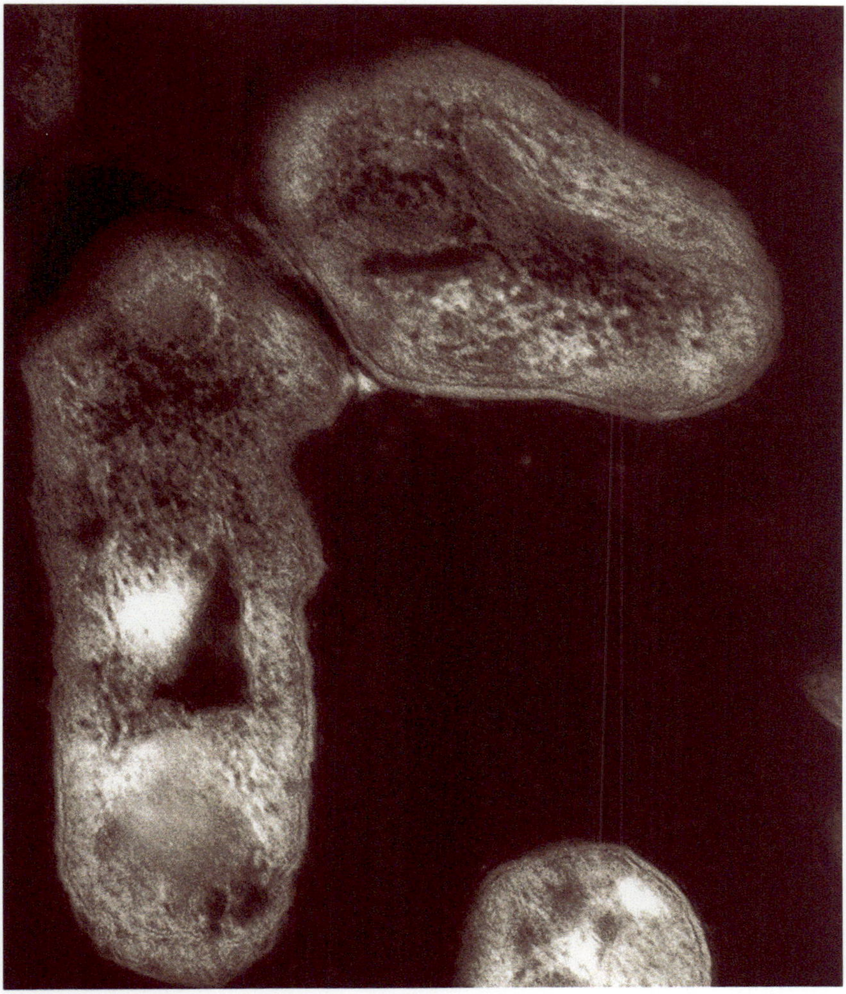

FIGURE 4.36 Asymmetric constriction and division in *Mycobacterium tuberculosis.* Bacilli had asymmetric constriction and division with unusual deviation in the position of constriction, in the typical "snapping postfission," to give a shorter daughter cell and a long daughter cell.

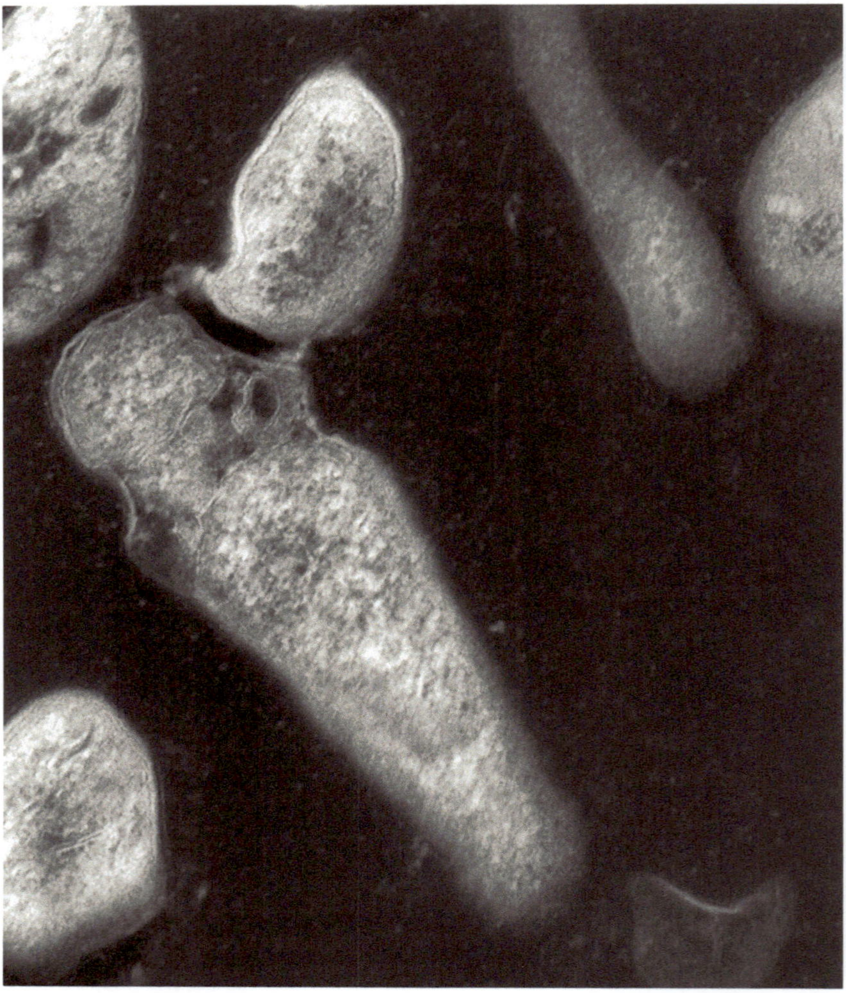

FIGURE 4.37 Asymmetric constriction in *Mycobacterium tuberculosis*.

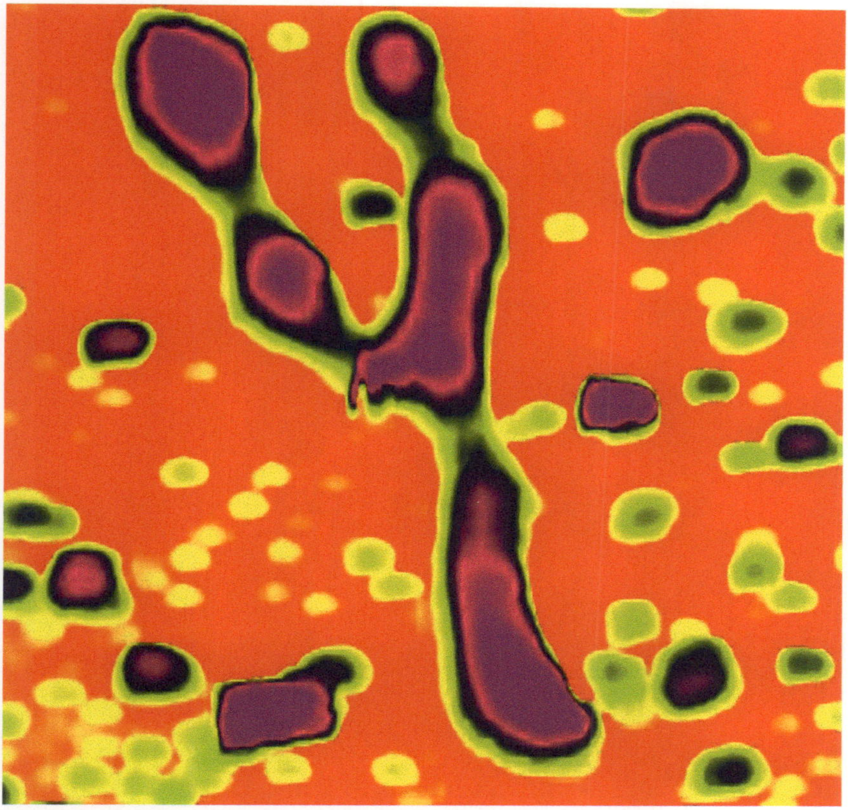

FIGURE 4.38 The totally drug-resistant *Mycobacterium tuberculosis* is known to form Y-shaped cells with branches more interior to the cells. These cells have fully developed dense bodies in polar positions, which may indicate they have cytoplasmic and nuclear content.

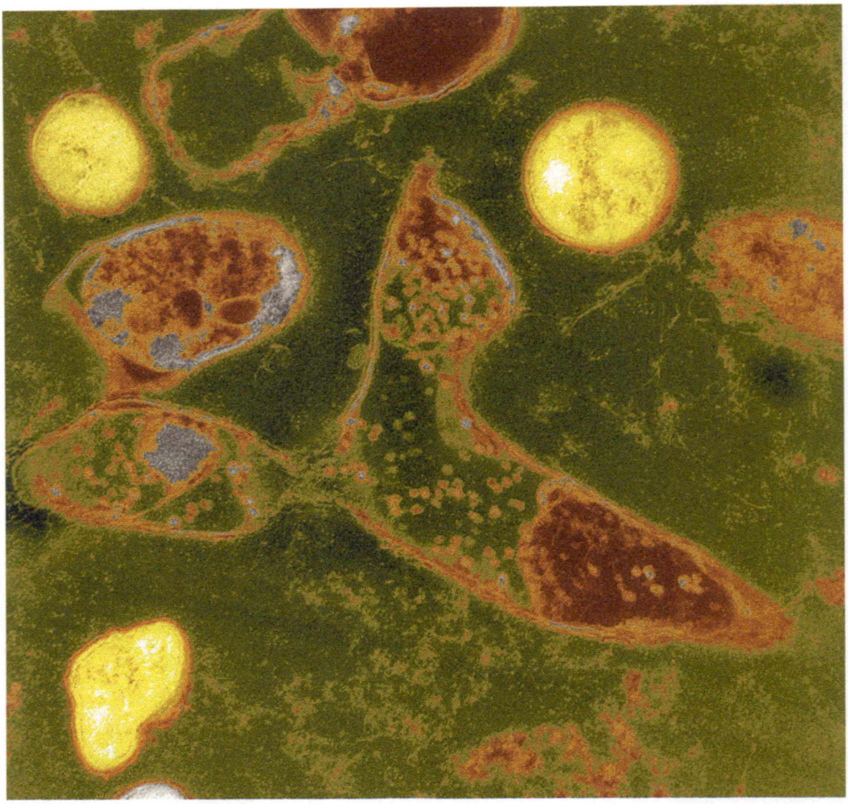

FIGURE 4.39 Y-shaped *Mycobacterium tuberculosis* can break down into masses of short rods or round cells.

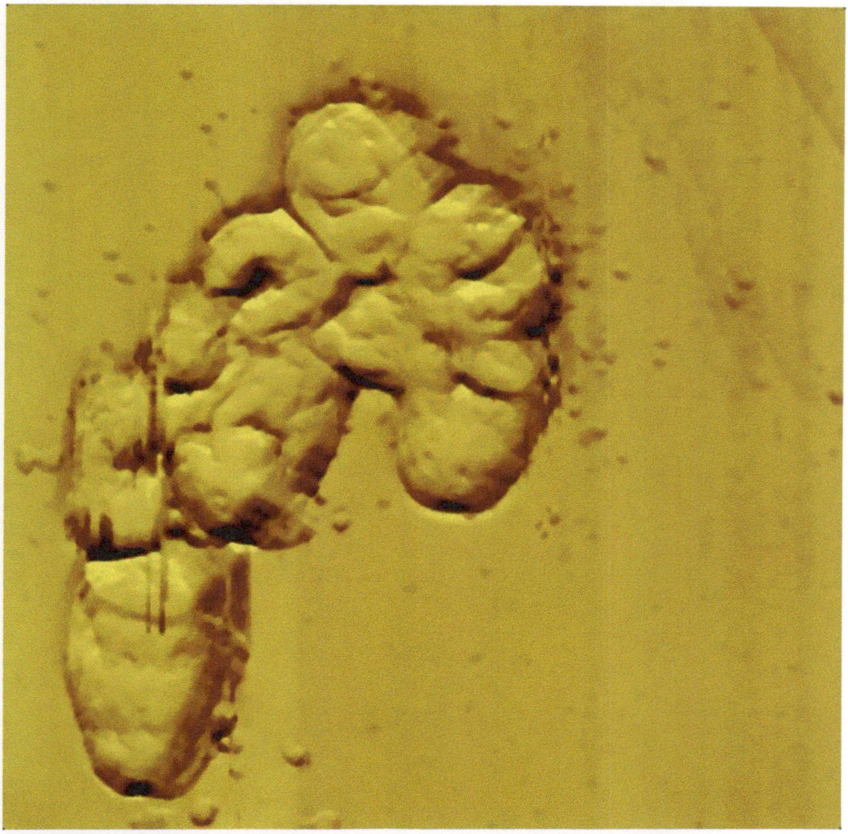

FIGURE 4.40 Budding in *Mycobacterium tuberculosis* recorded by the friction mode of atomic force microscopy. It was shown that budding does occur in highly drug-resistant strains of *Mycobacterium tuberculosis*. These cells are different from small buds that do not grow to any appreciable size.

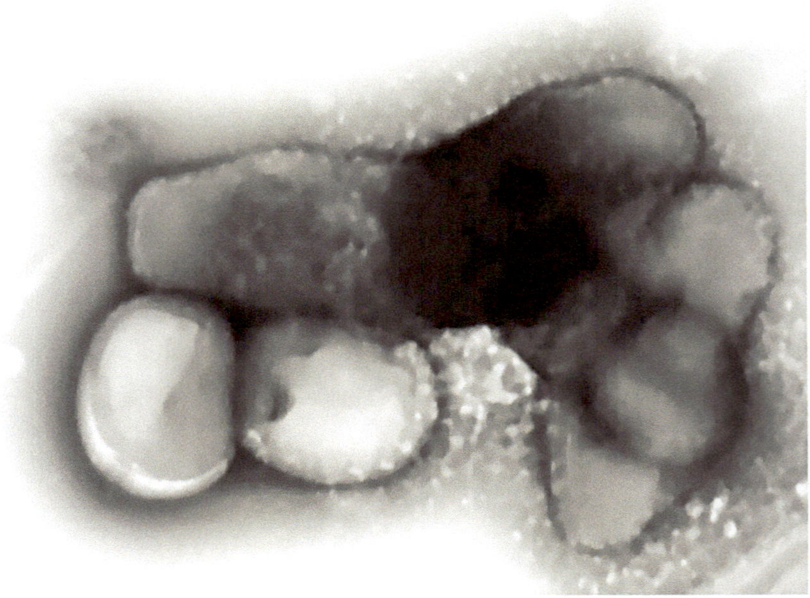

FIGURE 4.41 Budding in highly drug-resistant strains of *Mycobacterium tuberculosis* recorded by transmission electron microscopy.

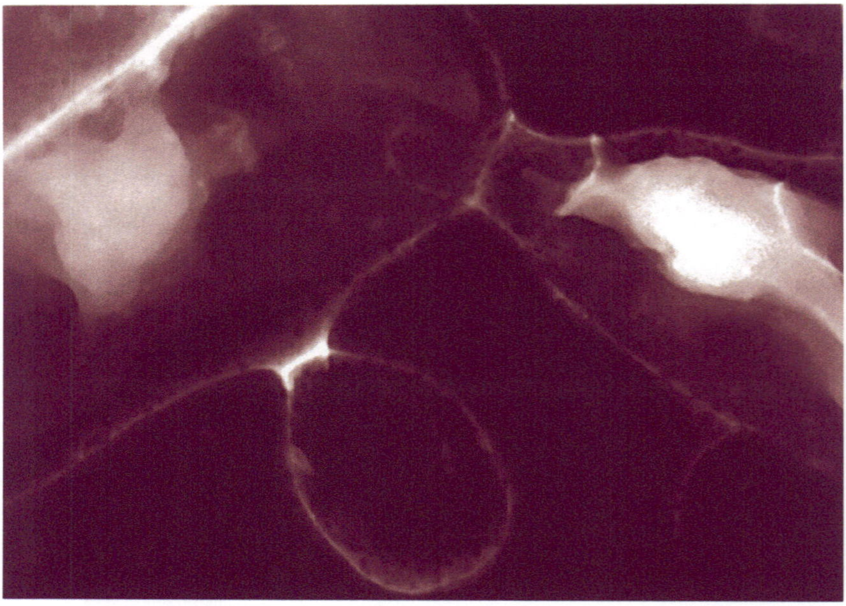

FIGURE 4.42 Side-wall budding in highly drug-resistant strains of *Mycobacterium tuberculosis.* In general, buds form at the distal end of *Mycobacterium tuberculosis* cells, but they also can be seen on the side wall of cells.

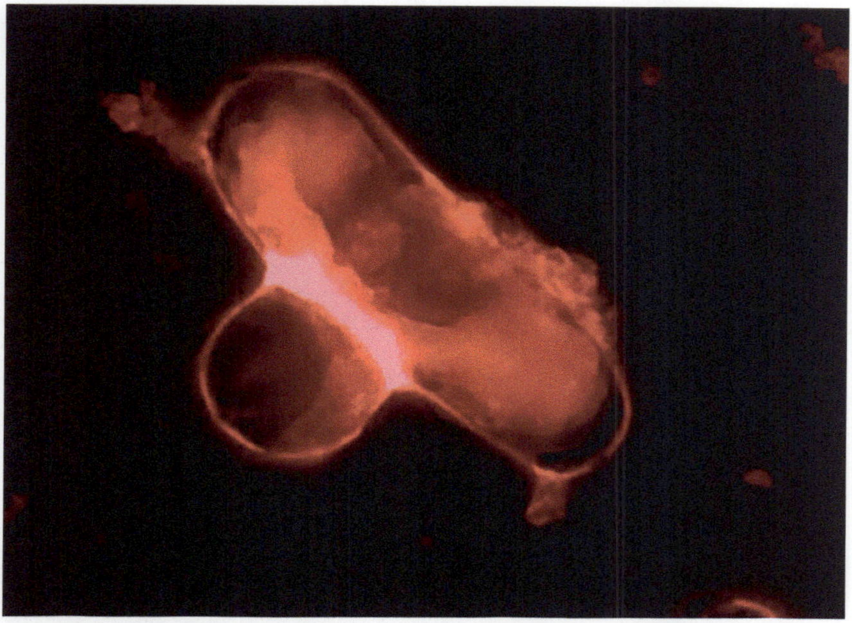

FIGURE 4.43 Budding in *Mycobacterium tuberculosis* recorded by transmission electron microscopy.

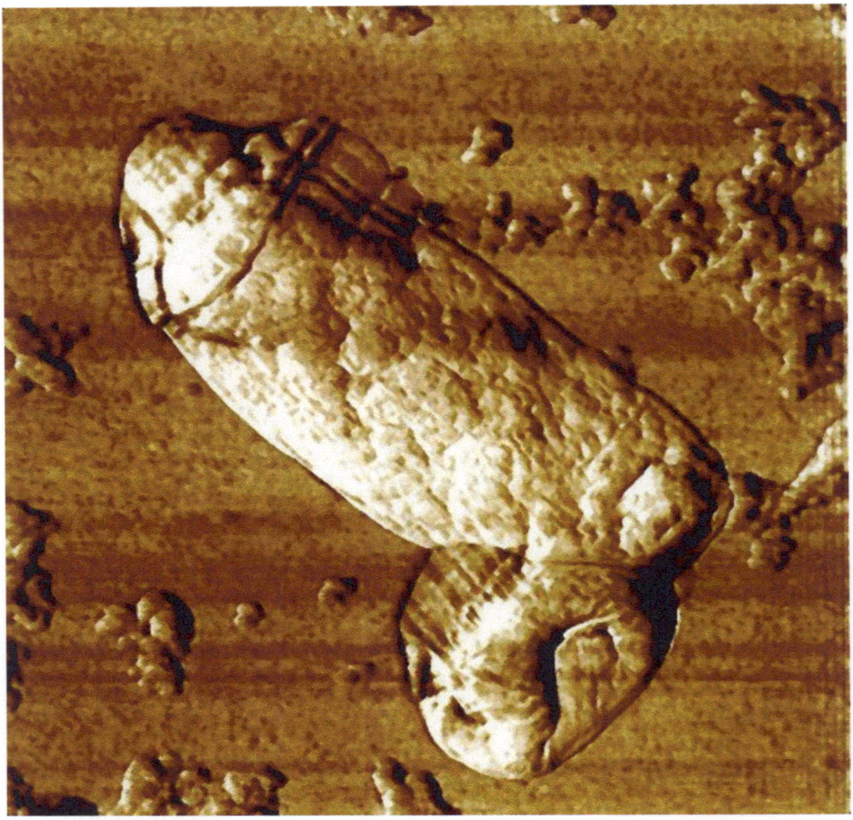

FIGURE 4.44 Atomic force microscopy recorded budding from the distal end of *Mycobacterium tuberculosis.*

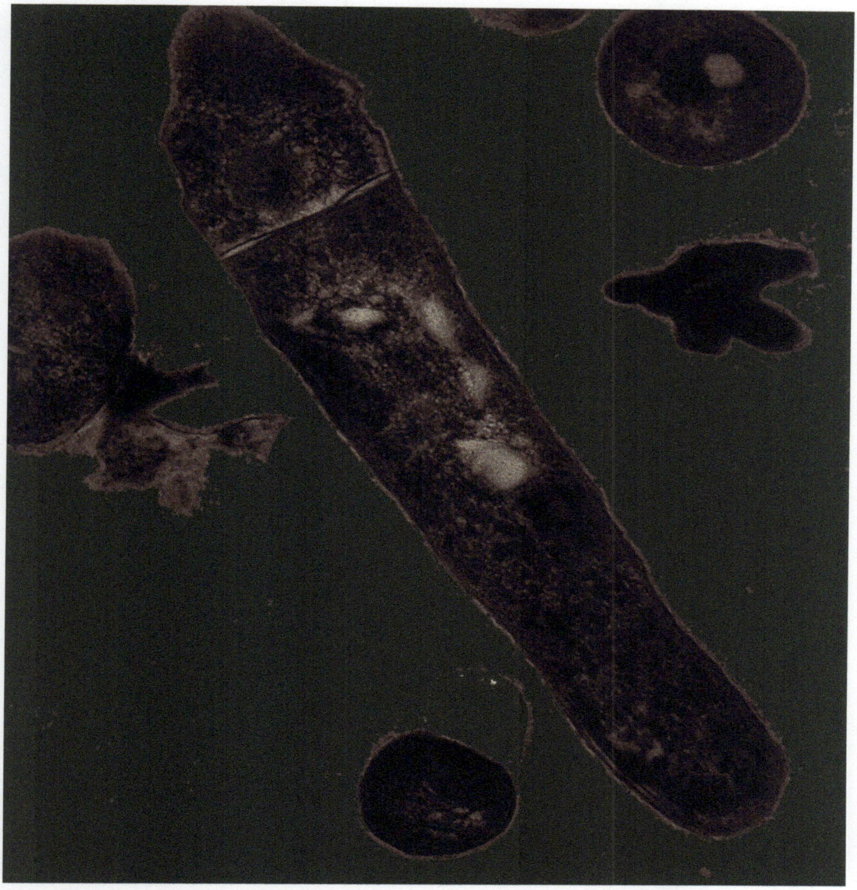

FIGURE 4.45 Asymmetrical septum in cell division. The cell formed by the asymmetrical type of division is different from buds that formed in highly drug-resistant strains of *Mycobacterium tuberculosis.*

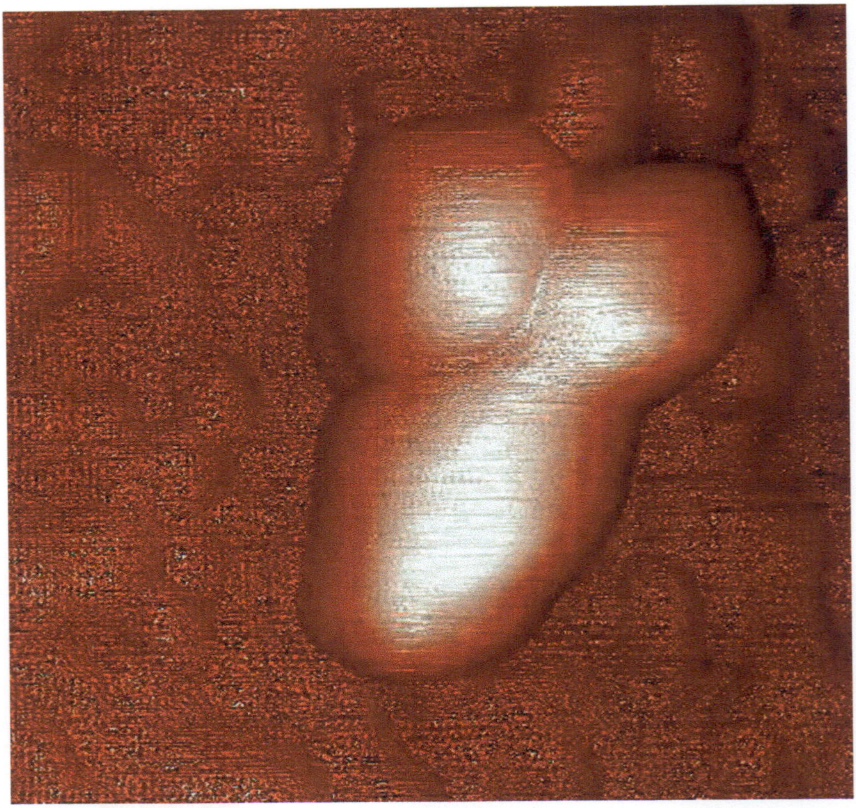

FIGURE 4.46 The small rod-shaped highly drug-resistant strains of *Mycobacterium tuberculosis* (0.5–0.6 μm). These cells are capable of dividing and/or branching.

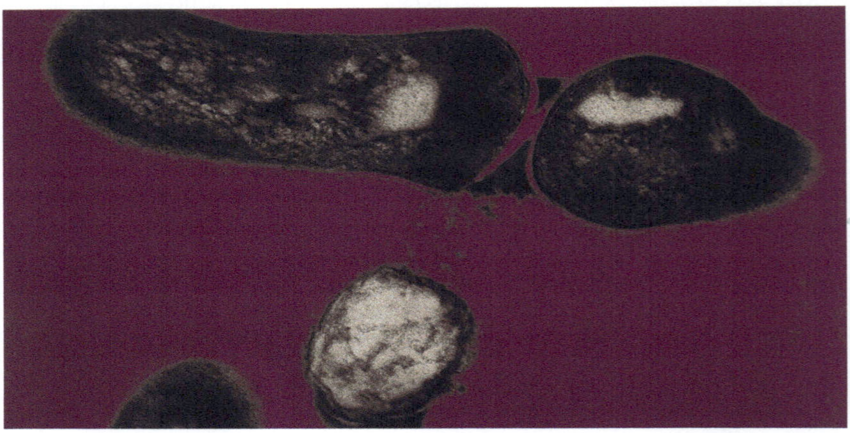

FIGURE 4.47 Asymmetrical type of cell division in *Mycobacterium tuberculosis* produces daughter cells of unequal size.

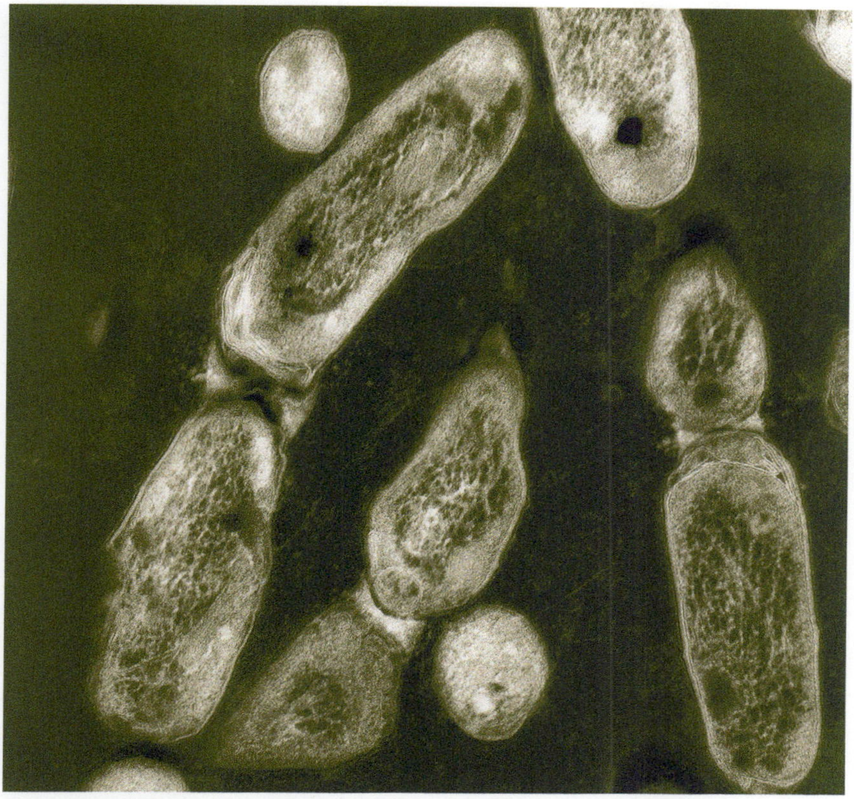

FIGURE 4.48 Symmetrical and asymmetrical type of cell division in *Mycobacterium tuberculosis* in the exponential phase of growth.

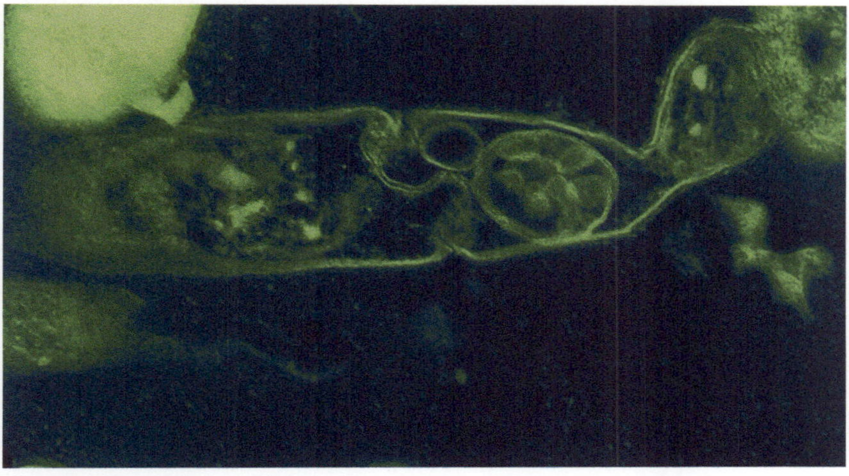

FIGURE 4.49 Asymmetrical type of cell division in heat-killed *Mycobacterium tuberculosis* recorded by transmission electron microscopy.

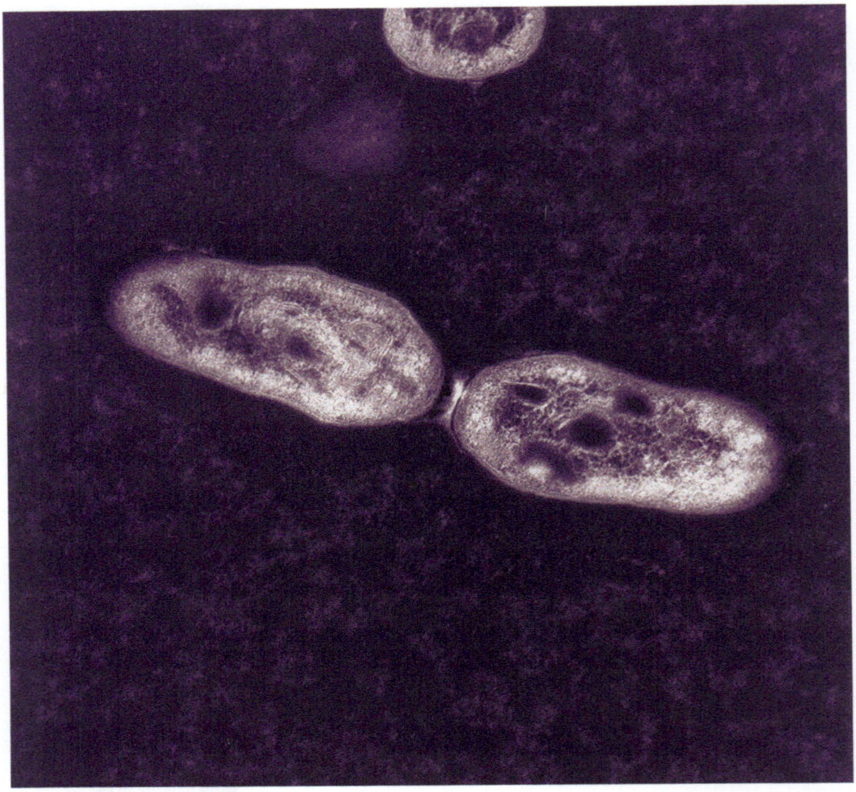

FIGURE 4.50 Symmetrical type of cell division in *Mycobacterium tuberculosis* recorded by transmission electron microscopy.

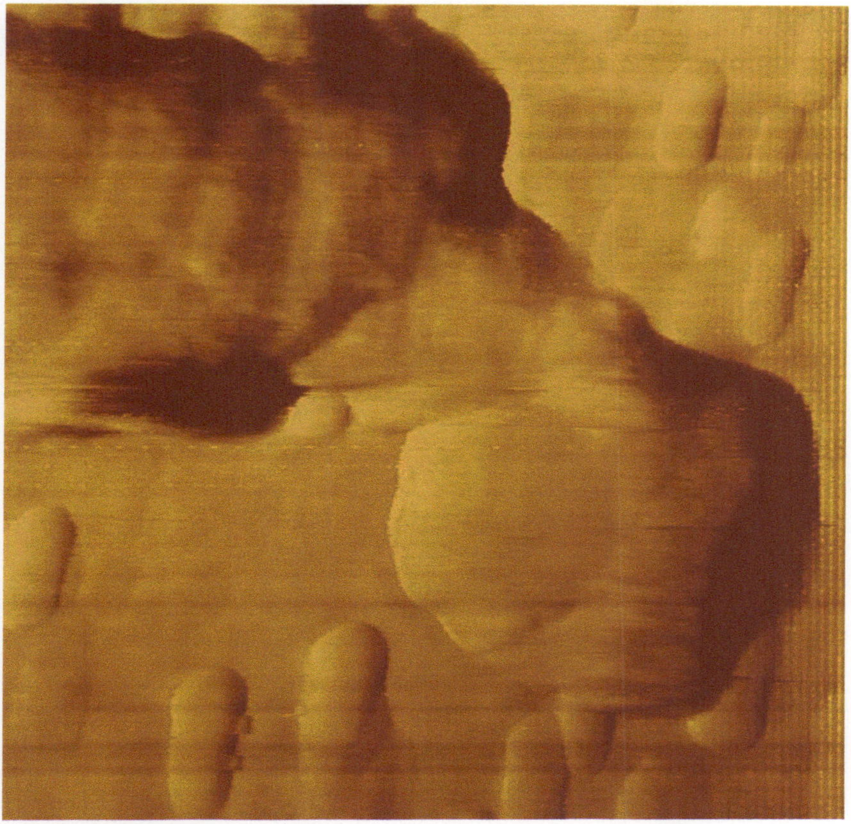

FIGURE 4.51 Formation of small round *Mycobacterium tuberculosis* bacilli from rod-shaped mother cells recorded by atomic force microscopy.

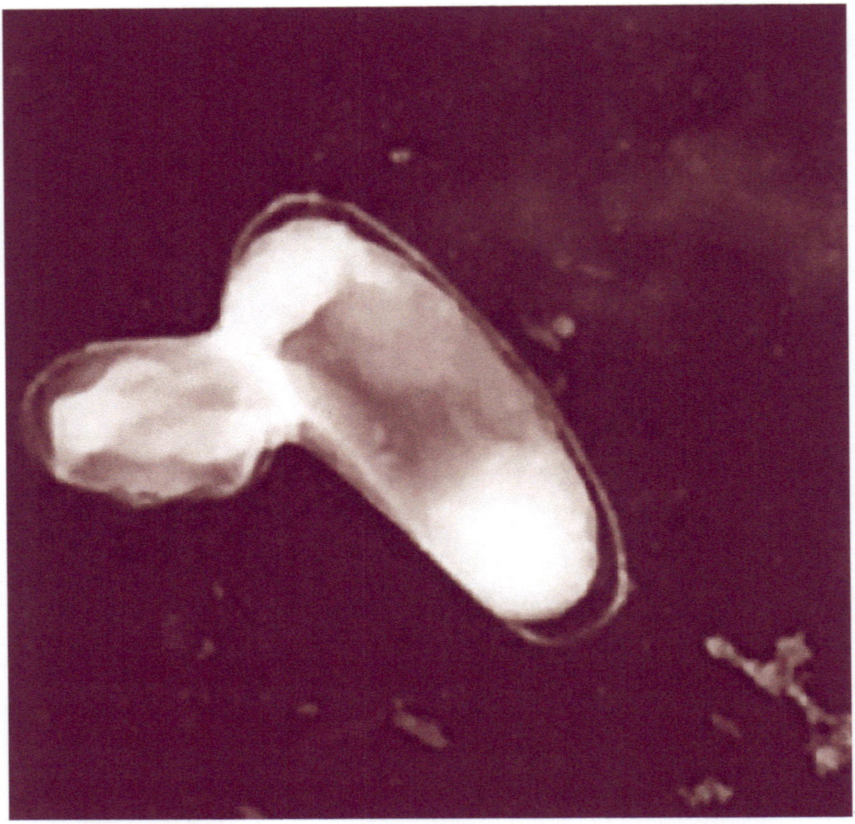

FIGURE 4.52 The budding type of cell division in highly drug-resistant *Mycobacterium tuberculosis* recorded by transmission electron microscopy.

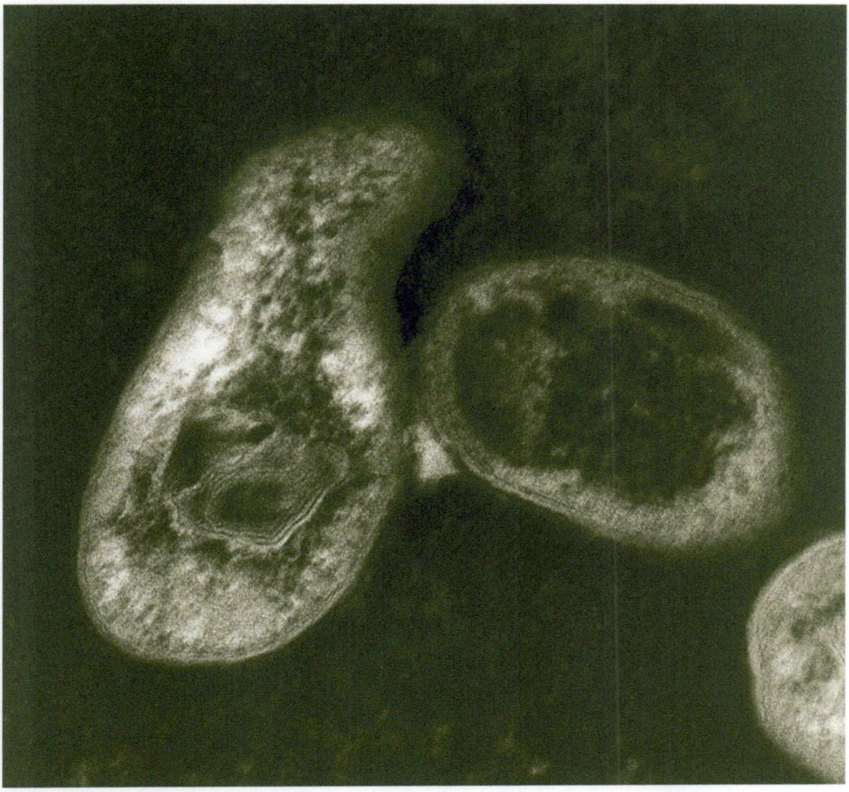

FIGURE 4.53 The longitudinal section of highly drug-resistant *Mycobacterium tuberculosis* that produces buds. It seems that the new progeny have enough of nuclear material and cytoplasmic content.

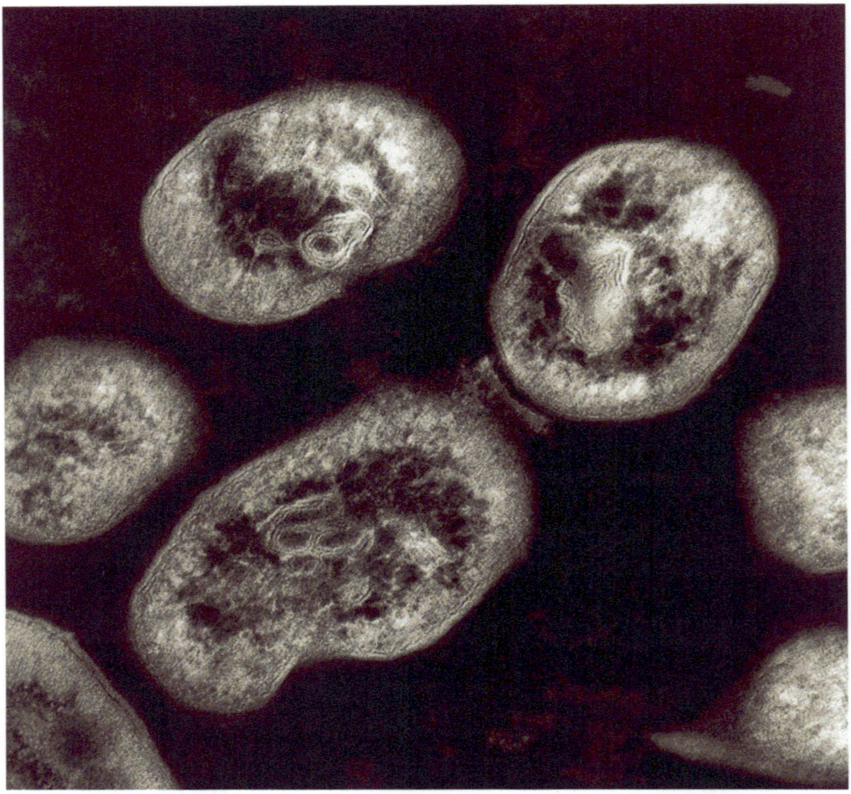

FIGURE 4.54 Asymmetrical type of cell division in highly drug-resistant *Mycobacterium tuberculosis* recorded by transmission electron microscopy.

Chapter 5

Drug-Resistant *Mycobacterium tuberculosis*

Tuberculosis (TB) remains a major cause of morbidity and mortality worldwide. According to the World Health Organization (WHO), one-third of the world population is infected with *Mycobacterium tuberculosis* and approximately 9 million new cases of TB occur each year. Despite the well-structured treatment programs and availability of effective treatment options since the 1950s (Fox et al., 1954; Mitchison and Selkon, 1957; Crofton, 1959; Mitchison, 1985; Santha et al., 1989; Da Silva and Palomino, 2011; Caminero et al., 2010), the estimated global TB mortality rate is close to 2 million (WHO, 2014). These death rates only partially depict the global TB threat because more than 70% of the TB patients are in the economically productive age of 15–50 years (Wejnert et al., 2015; Pealing et al., 2015). Among the associated risk factors, the HIV epidemic, underlying diseases, and decline of the socioeconomic were standard outlined to contribute to the disease's resurgence in different parts of the world (Barnes et al., 1991; Kim et al., 1995; Wang et al., 2005; Kang et al., 2014). Furthermore, the problem of drug resistance that was recognized in the very early days of the chemotherapeutic era (Pyle, 1947) raises a serious concern (Nachega and Chaisson, 2003). Although the resistance to a single agent is the most common type, resistance to multiple agents is less frequent but of greater concern (Frieden et al., 1993; Edlin et al., 1992; Fischl et al., 1992; Acosta et al., 2014; Caminero, 2015).

By convention, "multidrug resistance" (MDR) is defined as resistance to at least isoniazid (INH) and rifampin (RF) (CDC, 1990, 1991; Vareldzis et al., 1994; WHO, 1997, 2000; Martin and Portaels, 2007). WHO estimated that approximately 5% of all new TB cases of infections involved MDR-TB (WHO, 2014). Strains that combine MDR with additional resistance to at least one fluoroquinolone and one injectable second-line anti-TB drug have been appropriately named extensive drug-resistant tuberculosis (XDR-TB; CDC, 2006; Masjedi et al., 2006; Cohen, 2006; Mc Laughlin et al., 2007; Samper and Martín, 2007; Shah et al., 2007; Shean et al., 2008; Falzon et al., 2013). Recently, a more dangerous form of TB bacilli was identified, totally drug-resistant tuberculosis (TDR-TB) or extremely drug-resistant TB (Migliori et al., 2007; Velayati et al., 2009a, 2013; Udwadia et al., 2012; Klopper et al., 2013). These strains were resistant to all first- and second-line anti-TB drugs (Velayati et al., 2009a,b,c). Collectively, 9% and 2% of MDR-TB strains become XDR-TB and TDR-TB, respectively

(Velayati et al., 2009a, 2010, 2013; WHO, 2014; D'Ambrosio et al., 2015). However, this number may not reflect the correct situation because many laboratories in endemic TB countries do not have proper facilities and updated protocols to detect the XDR-TB or TDR-TB strains. Consequently, there is a stressing concern for additional control measures, such as new diagnostic methods, better drugs for treatment, and more effective vaccines (Weyer et al., 2013; WHO, 2014; Delogu et al., 2015; Andersen and Urdahl, 2015), to prevent the spread of these strains around the world.

The drug resistance to *M. tuberculosis* results from spontaneous and random mutations in the bacterial chromosome that results in reduced susceptibility to specific agents (David, 1970; Rastogi and David, 1993; Fox et al., 1999; Joshi, 2011). Although these mutations are rare events, a mutation early in multiplication produces a clone of resistant bacilli that are more frequently found. The incident of drug-resistant cases depends on the number of bacilli and the probability of drug-resistant mutants in the lesion. The mutation rates were 2.6×10^{-8} for INH and 2.2×10^{-10} for RF, whereas the more useful estimates of the highest proportion of mutants that can be expected in an unselected bacterial population were found to be 3.5×10^{-6} for INH and 3.3×10^{-8} for RF. For double drugs the rate is from 10^{-12} to 10^{-14} and 10^{-18} to 10^{-20} for three drugs (David, 1970; Crofton, 1970; Shimao, 1987; Grange, 1990; Davis, 1994; Colijn et al., 2011). Recent data have demonstrated that subpopulations of drug-tolerant persister cells are present even in growing- and stationary-phase cultures of *M. tuberculosis* (Keren et al., 2011; Sarathy et al., 2013a–c). This drug tolerance phenomenon is typical of TB and is partly attributed to the pathogen's ability to remain sequestered in macrophages and other stress-inducing microenvironments in a nonreplicating state of persistence (David, 1970; Mitchison, 1998; Chao and Rubin, 2010). There is no evidence that shows acquired genes or plasmids play a role in the emergence of antimicrobial resistance. However, the phenotypic and genotypic basis of drug resistance has been demonstrated (David, 1970; Zhang et al., 1992; Banerjee et al., 1994; Wilson and Collins, 1996). Therefore the natural drug resistance in *M. tuberculosis* is an important obstacle for the treatment and control of TB. This resistance has traditionally been attributed to the unusual multilayered cell envelope and active multidrug efflux pumps (Russell, 1969; Wayne et al., 1996; Wayne and Sohaskey, 2001; Sarathy et al., 2012, 2013a,c).

The cell envelope of *M. tuberculosis* is several-fold less permeable to chemotherapeutic agents when compared with functionally similar cell walls of other bacteria (Jarlier and Nikaido, 1994; Brennan and Nikaido, 1995; Ortalo-Magńe et al., 1995; Daffé and Draper, 1998; Daffe and Etienne, 1999). It is assumed that hydrophobic compounds more readily penetrate the cell walls by simply dissolving into and through the lipophilic cell wall, whereas the influx of hydrophilic compounds is largely facilitated by water-filled open channels that span the cell wall (Nikaido, 1994, 2003; Liu et al., 1996; Barry, 2001; Sarathy et al., 2013). In addition, *M. tuberculosis* has several adaptive mechanisms that increase the

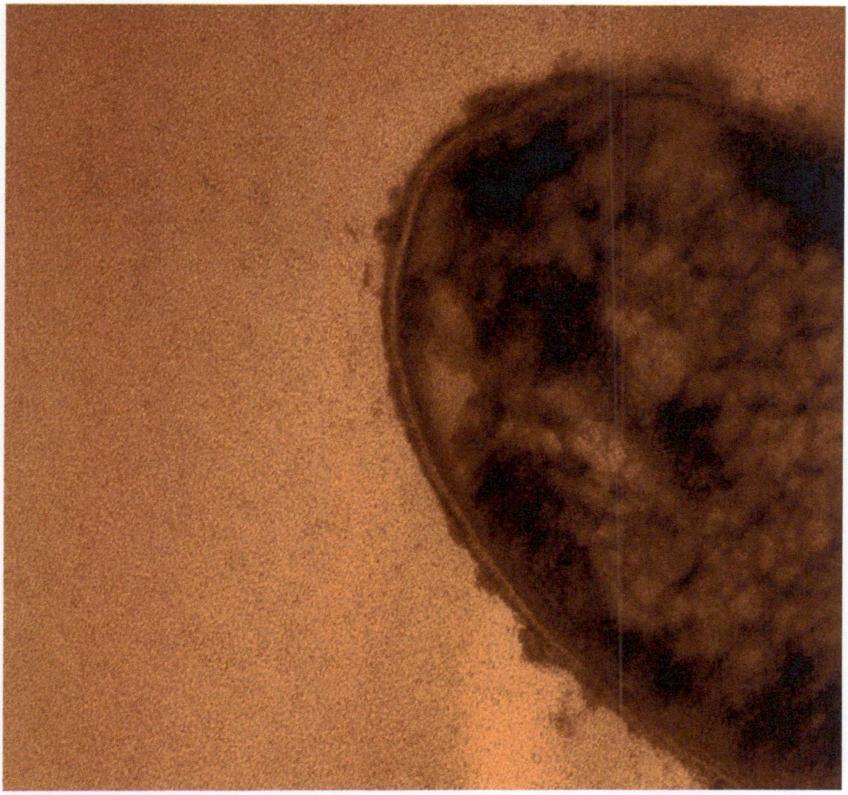

FIGURE 5.1 Cross-sections of susceptible *Mycobacterium tuberculosis*. The cell wall thickness is ~15.6±1.3 nm. The cell envelope is several-fold less permeable to chemotherapeutic agents compared with functionally similar cell walls of other bacteria.

level of their tolerance by preventing or minimizing their entry (Nikaido and Rosenberg, 1981; Nikaido, 2003; Sarathy et al., 2012). Drugs inhibiting these intrinsic systems would enable the entry of many antibiotics and would render *M. tuberculosis* more susceptible to various antibiotic classes (Martin and Portaels, 2007) (Figs. 5.1–5.16).

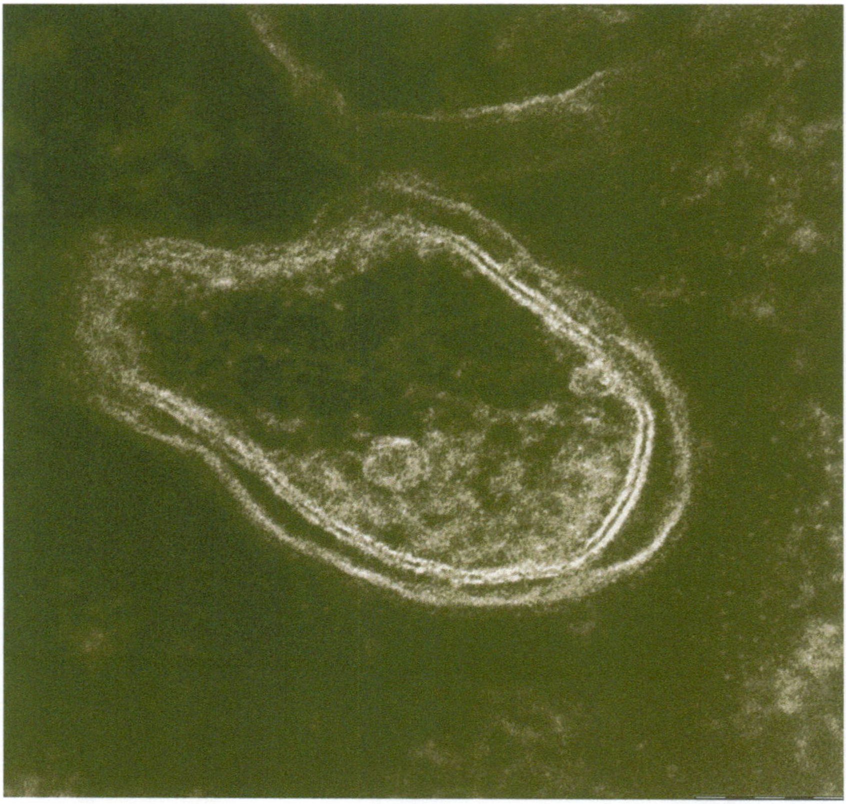

FIGURE 5.2 Longitudinal section from heat-killed susceptible *Mycobacterium tuberculosis*. The entire cell envelope can be divided into two main structural components: the cell membrane and cell wall. It is widely believed that the unusually high mycolic acid content, combined with various other intercalated lipids, contributes to the wall's limited permeability.

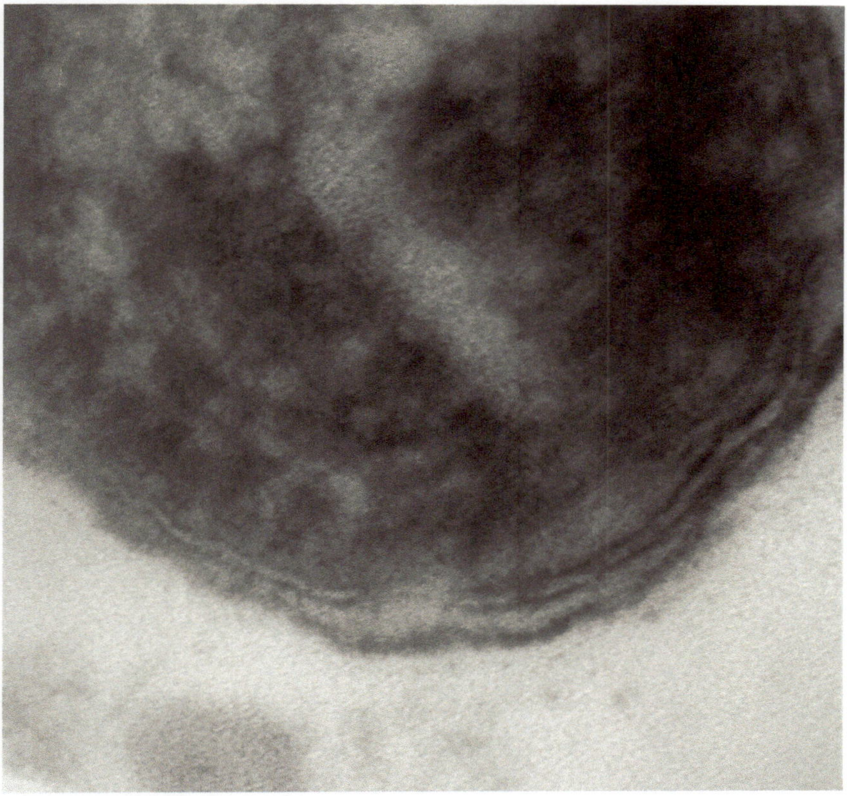

FIGURE 5.3 Cross-section of cell wall of *Mycobacterium tuberculosis*. The cell envelope is structurally distinct from that of gram-positive and gram-negative bacteria. The outer leaflet of the cell wall is composed of mycolic acids, which are covalently linked to the arabinogalactan–peptidoglycan complex of the inner leaflet.

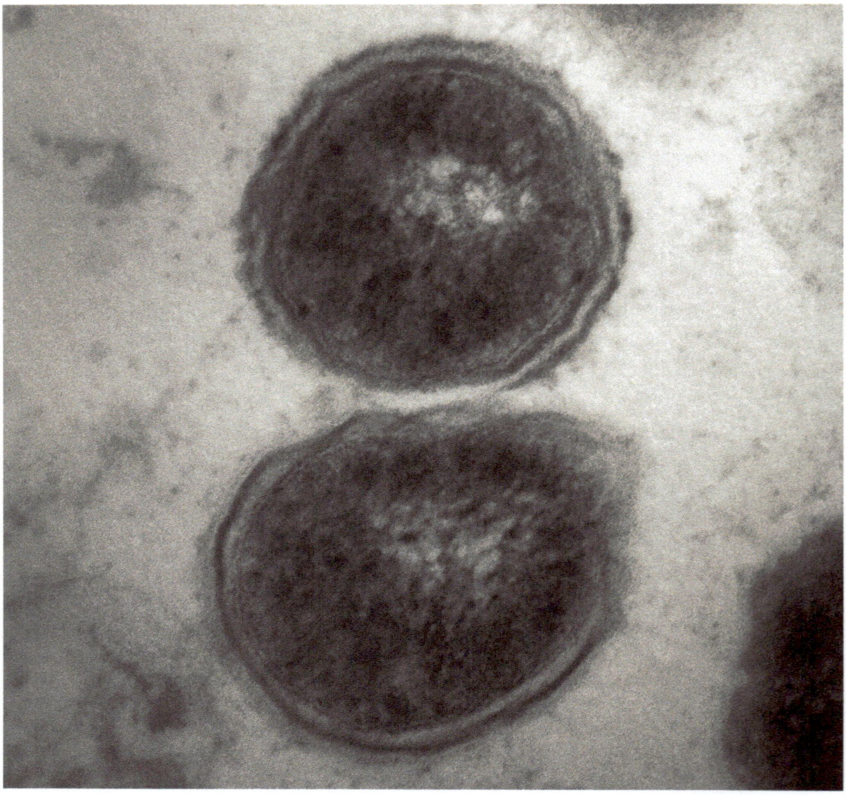

FIGURE 5.4 Cross-section of cell wall of multidrug-resistant strains of *Mycobacterium tuberculosis*. The thickness of the cell wall is ~17.1 ± 1.03 nm. Mycobacteria are capable of producing a multitude of mycolic acids with varying lengths and modifications depending on the species, strain, and growth conditions.

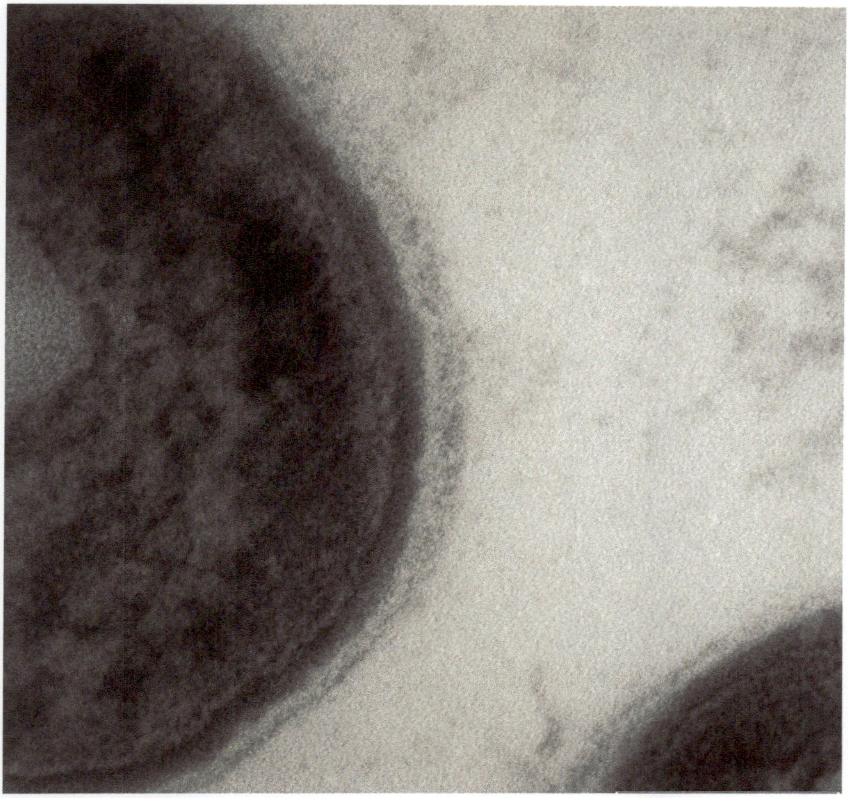

FIGURE 5.5 Cross-section of cell wall of multidrug-resistant strains of *Mycobacterium tuberculosis*. The differences in cell wall thickness may be attributed to the mycolate-to-peptidoglycan ratio, which needs to be investigated.

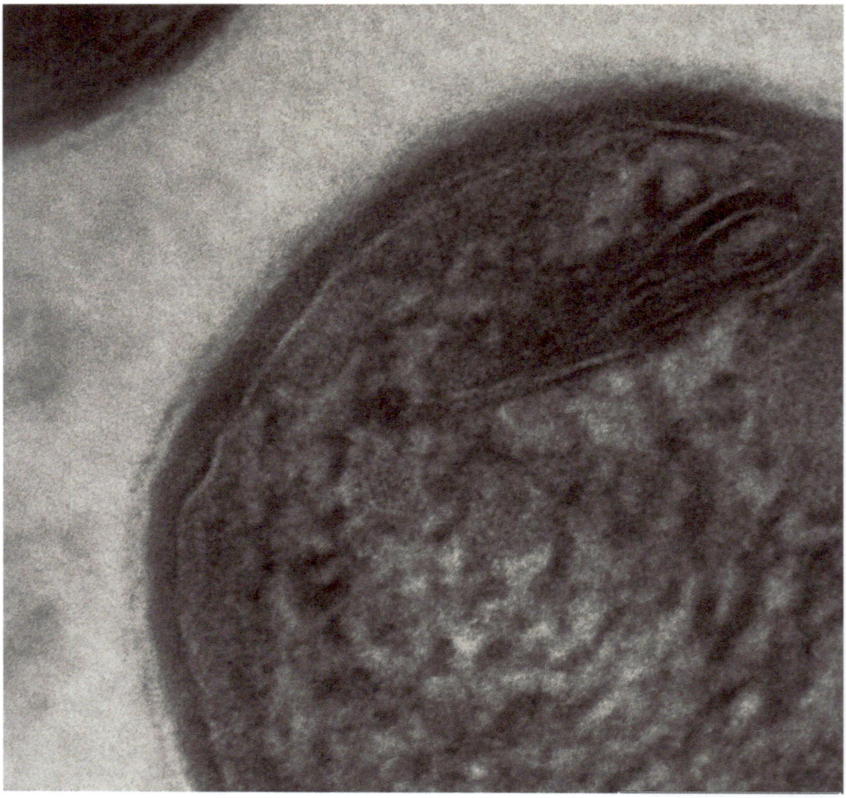

FIGURE 5.6 Cross-sections of the cell wall of extensively drug-resistant tuberculosis bacilli. The basal peptidoglycan layer is more electrons dense and in some cases was almost fused with the intermediate electron-transparent layer.

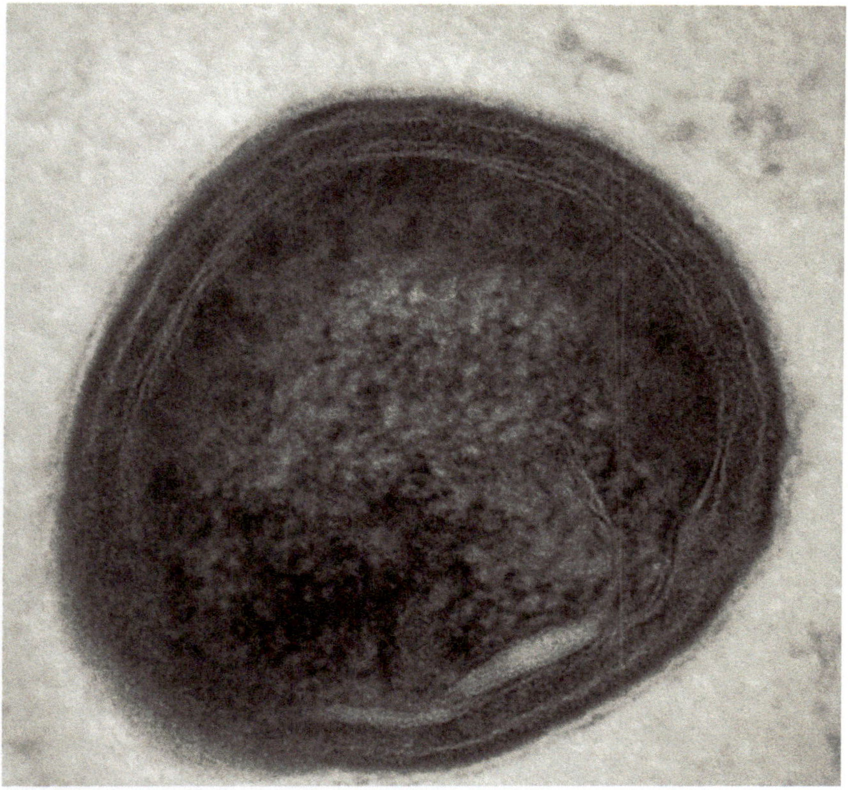

FIGURE 5.7 Cross-sections of the cell wall of extensively drug-resistant tuberculosis bacilli. The thickness of the peptidoglycan layer reaches 12–19 nm, whereas the same layer was 8–10 nm in susceptible cells.

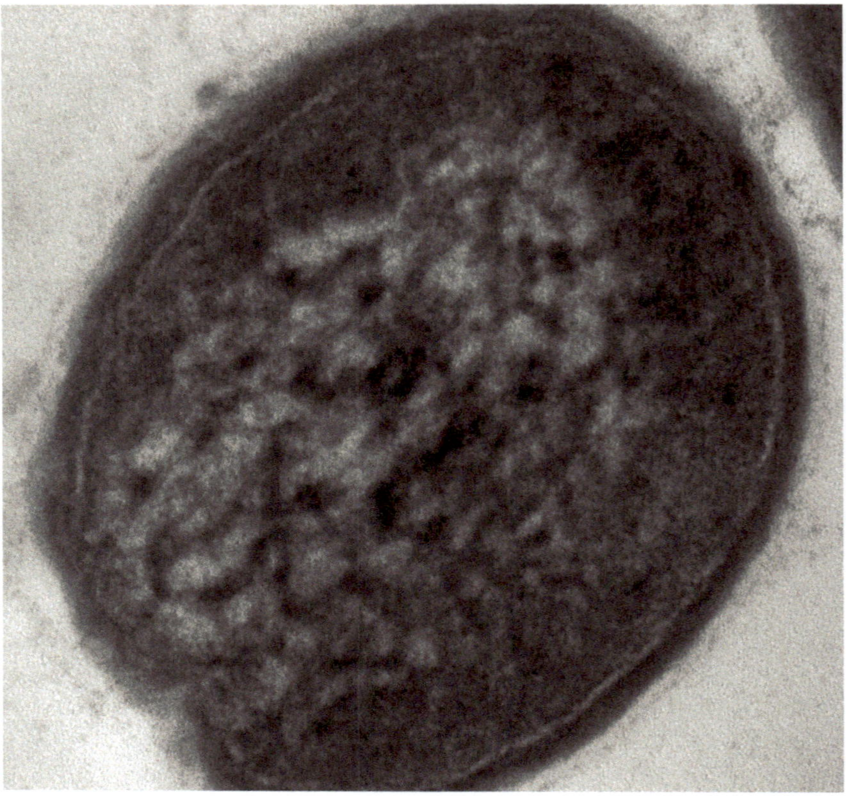

FIGURE 5.8 Cross-sections of the cell wall of extensively drug-resistant tuberculosis bacilli (XDR-TB). The XDR-TB is defined as strains resistant to at least one fluoroquinolone and one injectable second-line anti-tuberculosis drug in addition to isoniazid and rifampicin. Decreased cell wall permeability is suspected to influence fluoroquinolone resistance in *Mycobacterium tuberculosis*.

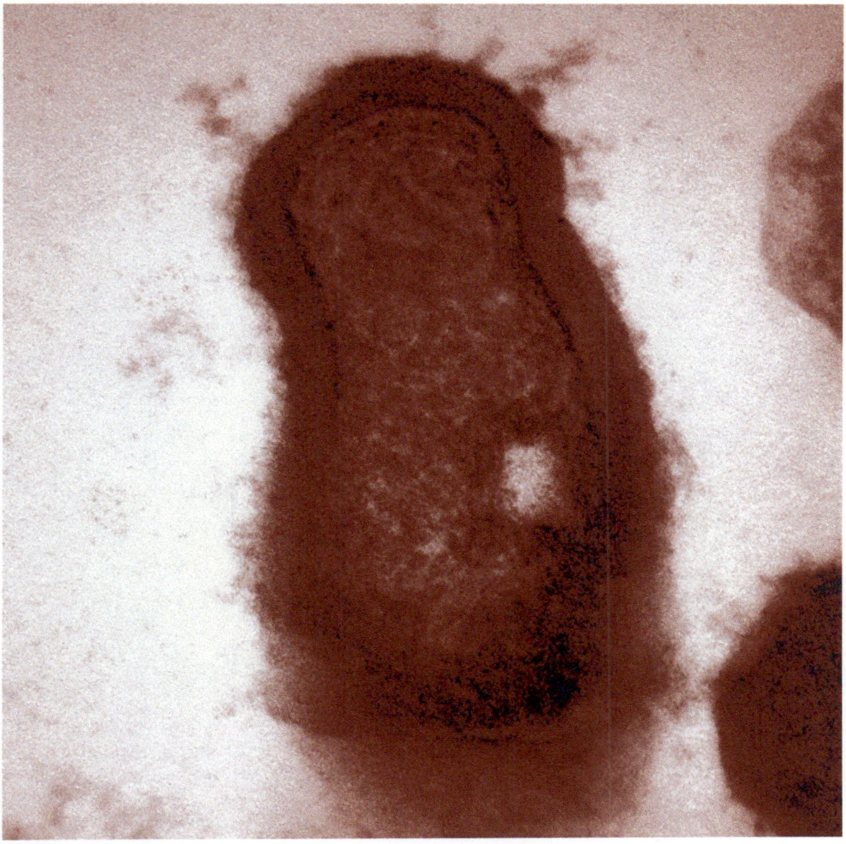

FIGURE 5.9 Transmission electron microscopy shows a totally drug-resistant tuberculosis bacilli. The cell wall thickness is observed in more than 50% of total populations. The cell wall thickness is variable from 20 ± 2.5 to 35 ± 1.5 nm.

FIGURE 5.10 Totally drug-resistant tuberculosis (TDR-TB) bacilli recorded by atomic force microscopy. The cell wall thickness is predominating in TDR-TB bacilli. This increase in cell wall thickness is believed to render them less susceptible than susceptible *Mycobacterium tuberculosis*.

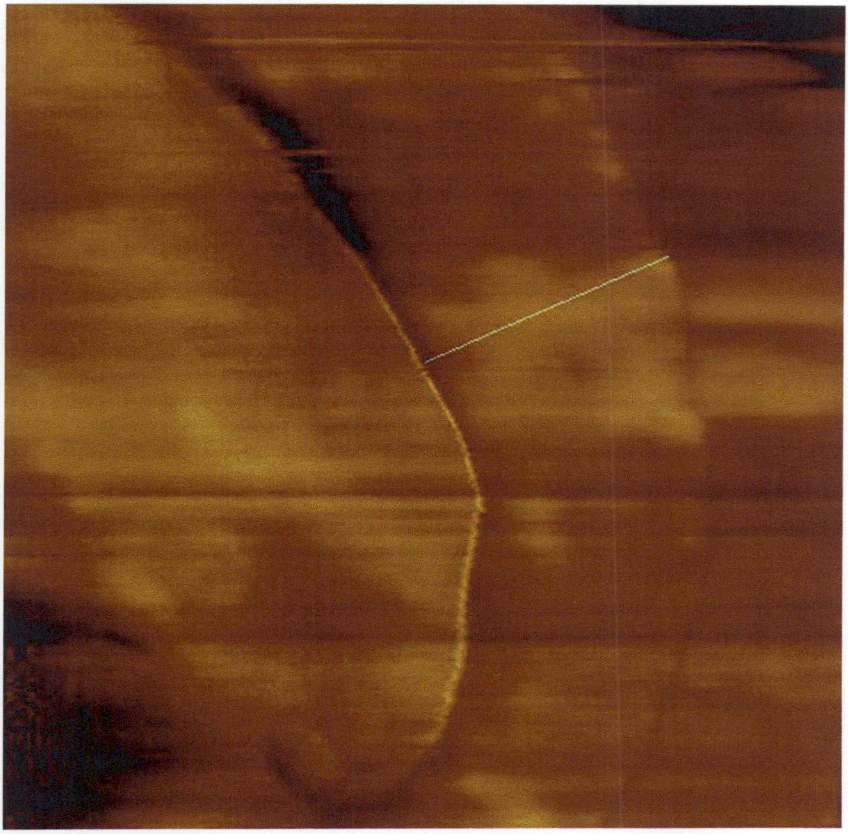

FIGURE 5.11 Cross-section of cell wall of totally drug-resistant tuberculosis bacilli recorded by atomic force microcopy (AFM). The cell wall thickness is clearer with AFM than transmission electron microscopy (TEM). Because interpreting TEM images requires more caution because electron microscopy analyses of ultrathin sections are performed with specimens from which water has been removed, water-soluble molecules may consequently aggregate and lipid molecules may be prone to extraction or rearrangement by organic solvents during dehydration. However, the cross-section using AFM was performed in water, and there is no need to dehydrate the samples. Under AFM the cell wall thickness reaches 35 ± 1.5 nm, almost triple what is usually present in susceptible forms.

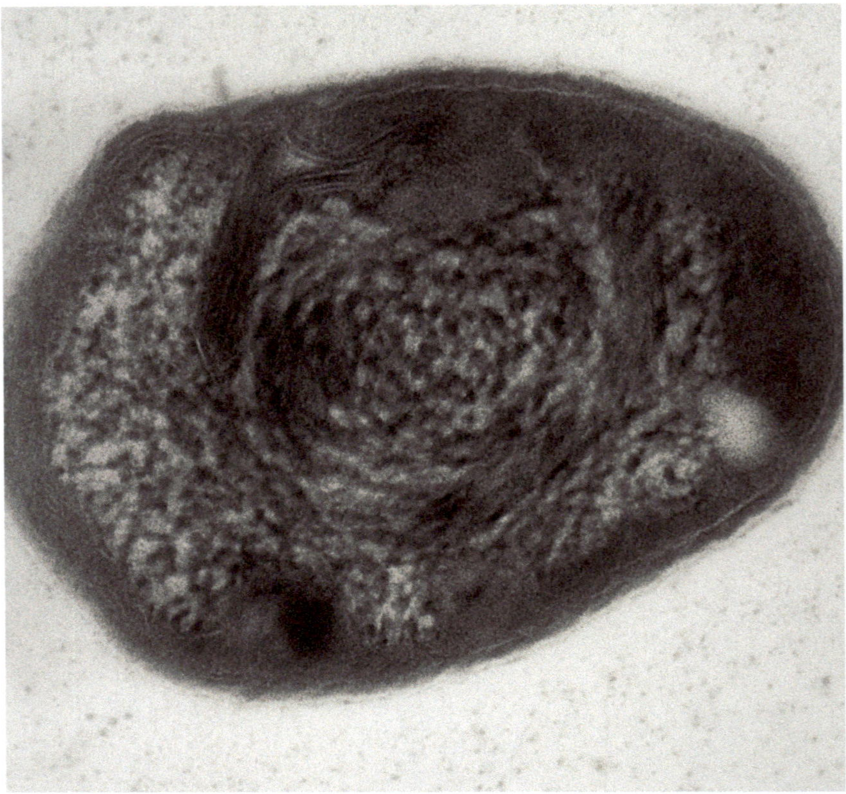

FIGURE 5.12 Longitudinal section of totally drug-resistant tuberculosis bacilli.

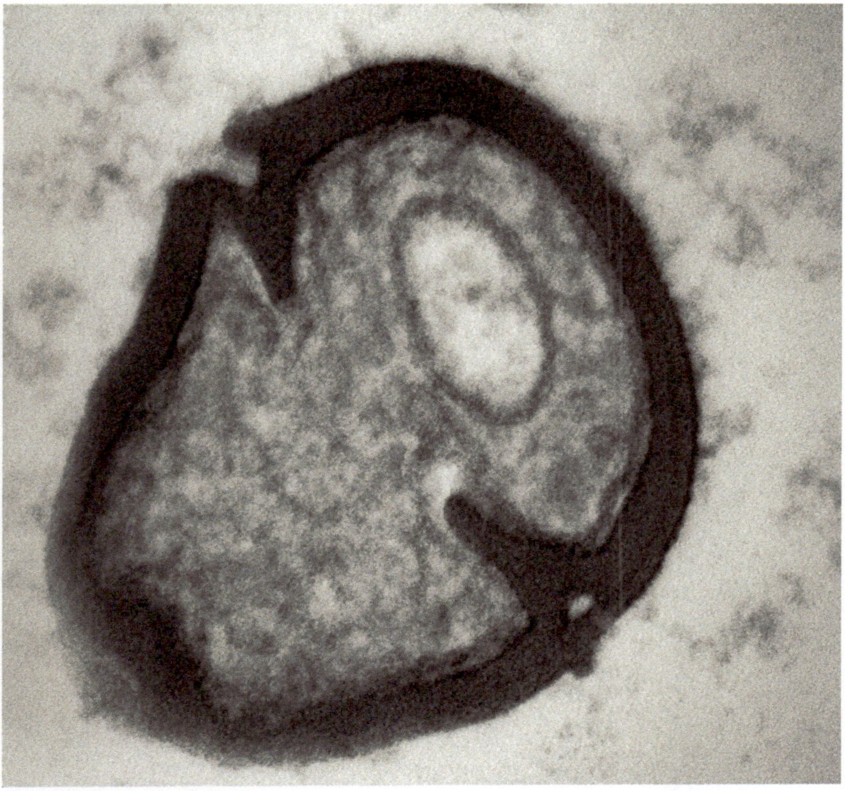

FIGURE 5.13 Longitudinal section of ovoid-shaped totally drug-resistant tuberculosis bacilli under transmission electron microscopy. The cell wall thickness is variable from 20 ± 2.5 to 35 ± 1.5 nm.

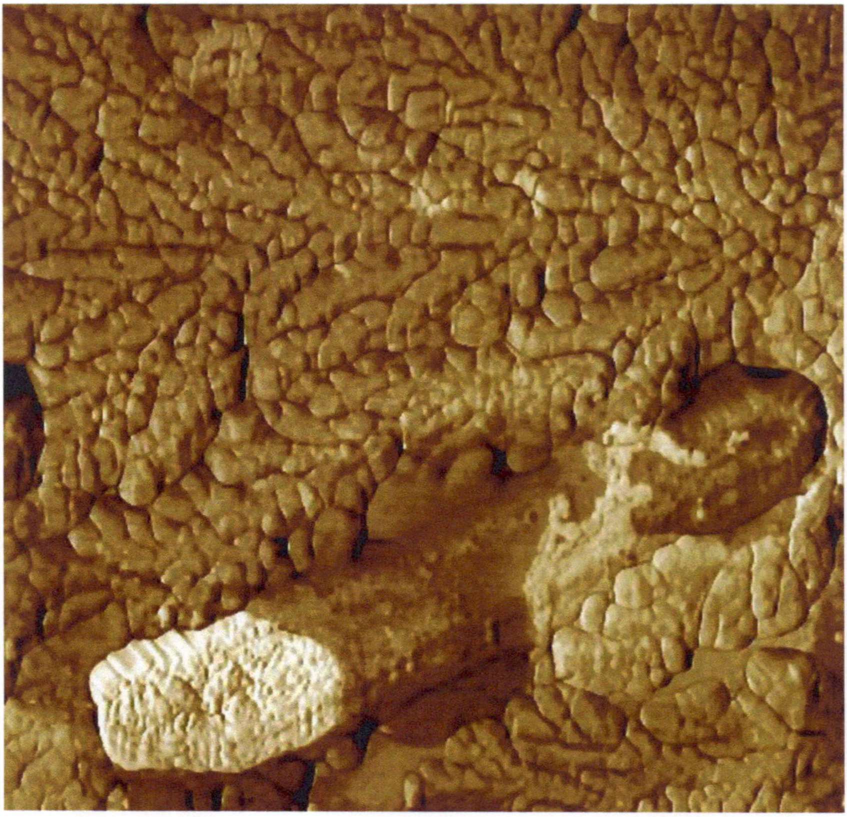

FIGURE 5.14 Totally drug-resistant tuberculosis (TDR-TB) bacilli recorded by atomic force microscopy. The TDR-TB bacilli were investigated after 72 h of co-culturing the bacilli in a cocktail of first- and second-line tuberculosis drugs. The cell wall damage is partial.

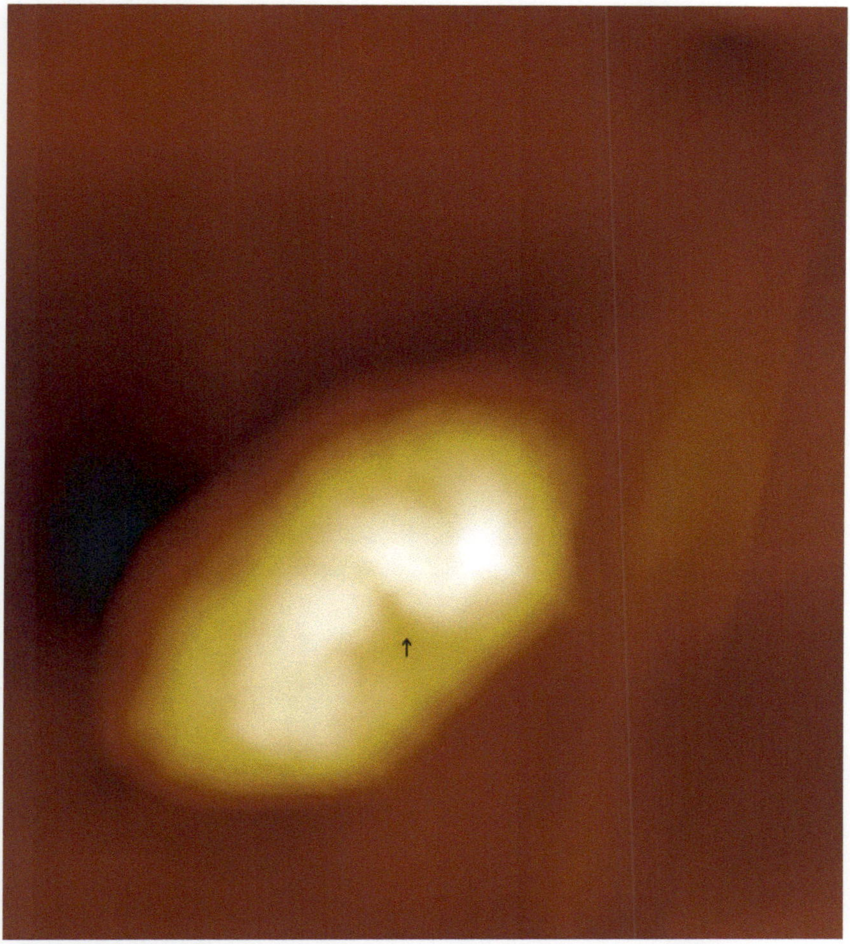

FIGURE 5.15 Ovoid-shaped tuberculosis-resistant bacilli. It has been shown that in *Mycobacterium tuberculosis* several pathways exist for compounds to cross the cell envelope. Among them are porins on the surface, which facilitate the influx of hydrophilic compounds.

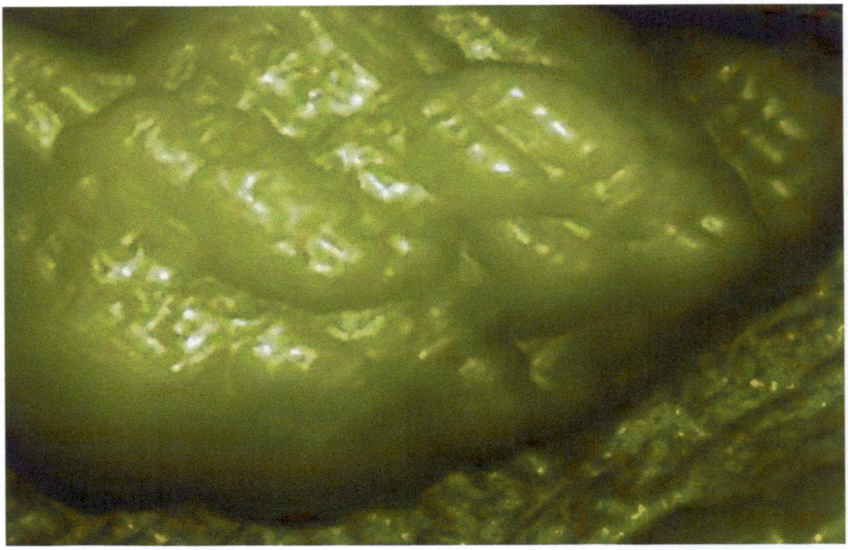

FIGURE 5.16 The cell surface of totally drug-resistant tuberculosis bacilli recorded by atomic force microscopy.

Chapter 6

Latency in *Mycobacterium tuberculosis*

Mycobacterium tuberculosis can persist within the human host for years without causing diseases in a syndrome known as latent tuberculosis (LTB; Opie and Aronson, 1927; Amberson, 1938; Feldman, 1938; Feldman and Baggenstoss, 1939; Sweany et al., 1943; Harada, 1976; Dannenberg, 1993; Hernandez-Pando et al., 2000; Lillebaek et al., 2002). The shift of *M. tuberculosis* into a persistence state can occur after initial infection (preantibiotic persistence) or after completion of TB treatment in TB patients who developed late reactivation (postantibiotic persistence; McDermott, 1958, 1959; Tomasz et al., 1970; Tuomanen, 1986; Khomenko, 1987; LeCoeur et al., 1989; Fox et al., 1999; Balasubramanian et al., 1994; Alland et al., 1998; Grange, 1998; Parrish et al., 1998; Michele et al., 1999; Wilson et al., 1999; Wallis et al., 1999; Flynn and Chan, 2001; Rieder, 2002; Cardona and Ruiz-Manzano, 2004; Gomez and McKinney, 2004; Cardona, 2007; Li et al., 2014).

Today more than one-third of the world's population (~2.2 billion individuals) is estimated to harbor LTB infection, of which 2–23% will develop active TB during their lifetime (WHO, 2008, 2013, 2014; Esmail et al., 2014). The reactivation process can be altered with HIV infection, malnutrition, drug use, cancer, diabetes, chronic renal insufficiency, and immunosuppressive drug therapy (Rastogi, 1993; Rastogi and Barrow, 1994; Parrish et al., 1998; Flynn and Chan, 2001; Ruslami et al., 2010; Ottmani et al., 2010; Esmail et al., 2014). Therefore more efforts must be devoted to better control their progression and to better understand the latent state. In the last 2 decades significant progress has been made to develop and characterize various aspects of LTB cells (Corper and Cohn, 1933; Wheate, 1951; American Trudeau Society, 1953; Reed, 1957; Vignal, 1882; Mitchison and Dickinson, 1978; Wayne, 1976; Wayne and Lin, 1982; Mochizuki, 1991; Morita, 1993; Kaprelyants et al., 1993; Wayne and Sramek, 1994; Mukamolova et al., 1998; Cunningham and Spreadbury, 1998; Michele and Bishai, 1999; Zhang et al., 2001; Mostowy and Behr, 2002; Shleeva et al., 2002; Betts et al., 2002; Gengenbacher et al., 2010; Chao and Rubin, 2010; Shleeva et al., 2011). After the introduction of TB chemotherapy in the late 1940s, surgical resection of tuberculous lesions became increasingly common and provided material for numerous bacteriological studies. Several reports in the mid-1950s showed that many acid-fact bacilli could be observed by microscopy in TB patients on chemotherapy (Beck and Yegian, 1952; Medlar et al., 1952; ATS, 1953; Hobby et al., 1954; Loring and Vandiviere, 1956; Wayne, 1956). However, most of these lesions failed to yield culturable bacilli; this was especially true of "close lesions" lacking patent bronchial communication. However, whether these visible but nonculturable bacteria were truly and irreversibly dead or were instead "viable but nonculturable" remained the subject of vigorous debate.

The Cornell model of latency was the first experimental evidence of the existence of latent bacilli in an in vivo model, and it is widely considered to be the experimental source of latent bacilli. The model was developed by McDermott and his colleagues at Cornell University in the 1950s. In this model, outbred mice were infected intravenously with approximately 10^5 colony-forming units of laboratory strains of *M. tuberculosis*. They were immediately treated for a period of 12 weeks with the antimycobacterial drugs isoniazid and pyrazinamide. For a 4- to 6-week period after withdrawal, the mice showed no evidence of culturable tubercle bacilli. This is termed the "sterile" state of LTB. However, after 12 weeks the culturable bacilli were recovered in two-thirds of infected mice. In fact, it became clear that the drugs had not eradicated the infection but merely rendered it "latent" in the sense that the presence of the infection cannot be demonstrated by any of the available methods and its presence can only be detected in appearance as relapse (ie, reappearance of culturable bacilli in the tissues; McCune and Tompsett, 1956; McCune et al., 1956, 1966). The rate of relapse could be increased to 100% with 1 mg/day^{-1} of cortisone for 20 days. Soon it was demonstrated that rifampicin could sterilize lesions in a remarkably short period of time compared with other drugs (Waksman, 1964; Mitchison, 1979, 1985).

Later on, based on the speed of growth of bacilli, Mitchison and Dickinson (1978) proposed a bacillary population in TB lesions. Bacilli with a high metabolism were highly susceptible to chemotherapy, a medium speed of growth, or "spurts of growth" "were susceptible to pyrazinamide or rifampicin, but not to isoniazid, and finally a latent bacilli showed no metabolic activity and were not sensitive to chemotherapy" (Mitchison, 1979; Grosset, 1980; Fox et al., 1999; Cardona and Manzano, 2004). Collectively, in most in vivo studies the existence of a subpopulation of bacteria that are phenotypically tolerant to the effects of antibiotics was demonstrated. However, the conditions that trigger the tolerant state were not cleared. Pathological studies on human lesions led to the hypothesis that the development of potentially bacteriostatic tissue environments (eg, reduced oxygen tension, nutrient limitation, or acidic pH) could result in the acquisition of drug tolerance by nonreplicating bacteria (Mochizuki, 1991; Wayne and Hayes, 1996; Hu et al., 1998; Betts et al., 2002; Karakousis et al., 2008; Deb et al., 2009; Shleeva et al., 2011). This hypothesis was encouraged by the studies of Corper and Cohn (1933), who showed survival of 0.01% of clinical isolates at 37°C after 12 years in sealed cultures. Wayne (1976, 1981) conducted studies regarding the bacteriological state of *M. tuberculosis*. In his model, cultures of *M. tuberculosis* are subjected to gradual oxygen withdrawal by incubation in sealed containers with controlled agitation. It was shown that after a progressive introduction to a low oxygen pressure, *M. tuberculosis* changed its metabolism by enhancing some enzymes, mainly isocitrate lyase and glycine dehydrogenase, to generate a reduced nicotinamide adenine dinucleotide so as to obtain energy through a fermentative metabolism (Wayne and Lin, 1982). However, these experiments did not answer the question of how bacilli may survive the stress condition in activated macrophages (eg, a low pH and a high concentration of radical oxygen intermediates and radical nitrogen intermediates; Ulrich et al., 2004). Considering an in vitro model of latency, Kaprelyants et al. (1993) distinguished three major physiological

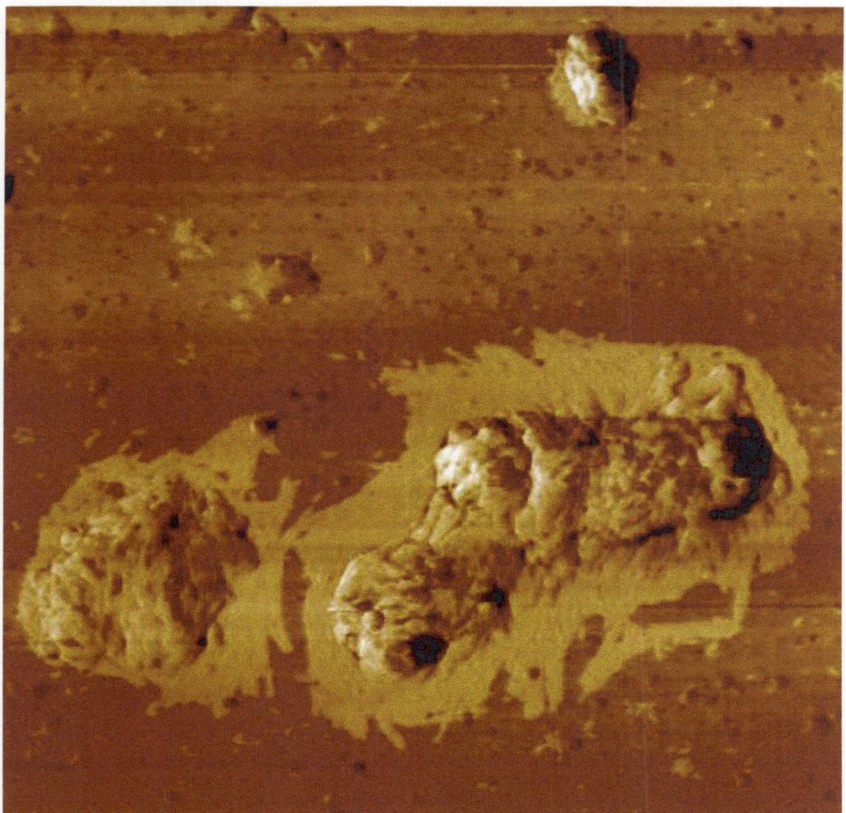

FIGURE 6.1 *Mycobacterium tuberculosis* after 2 weeks under stress conditions. The cell cultured under zero oxygen and low-nutrient Dubos medium (0.03% Tween and 10% Dubos albumin supplement) without glycerol and with stirring.

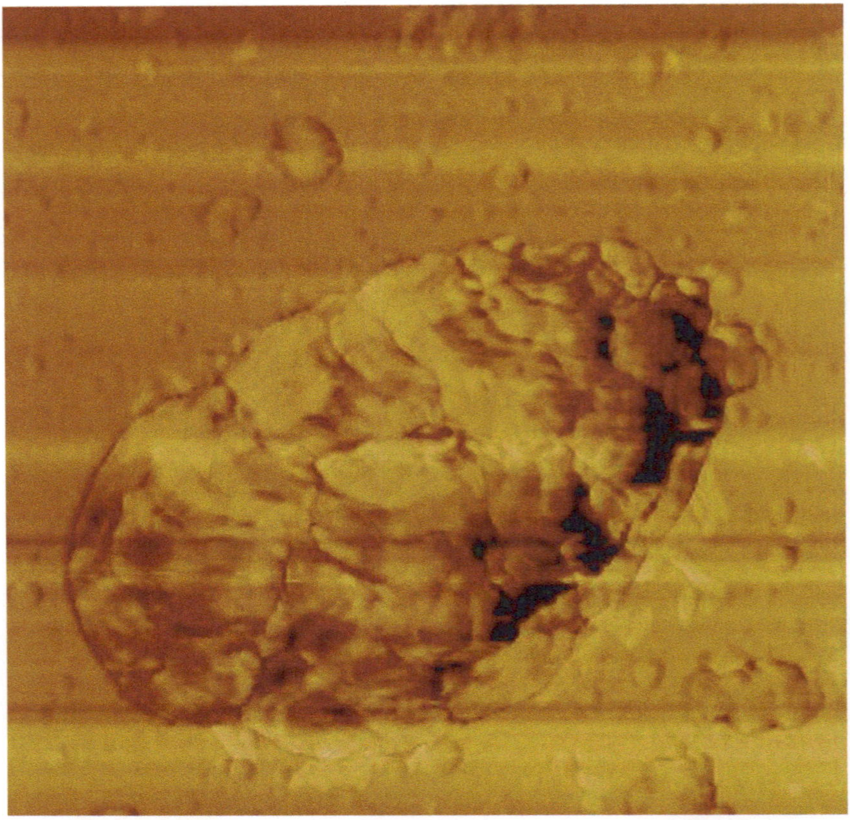

FIGURE 6.2 *Mycobacterium tuberculosis* after 1 month under stress conditions. *M. tuberculosis* developed cell wall thickening (21.5 ± 1.8 nm). The cell wall thickening may act as a protective coat and may offer protection against hostile environments by restricting the exchange of metabolic activity.

states for bacilli: (1) viable (culturable) bacilli that may divide, (2) dormant bacilli with a low metabolic activity that are unable to divide without a preceding resuscitation phase, and (3) nonviable (nonculturable) bacilli that cannot divide. However, it is difficult to extrapolate these data to the real circumstances experienced by the bacilli inside macrophages (Figs. 6.1–6.15).

The presence of nonacid-fast, cell-wall–defective variants of tubercle bacilli that was reported by Much (1907) became more evident. These cells were isolated from clinical specimens and mycobacterial cultures (Khomenko, 1987; Chandrasekhar and Ratnam, 1992). These forms seemed to be induced by the administration of antibiotics, as is the case with many other organisms (Mattman, 1970, 2000). In this respect a few investigators also proposed the ability of bacteria to exist as a population of self-replicating forms with defective or

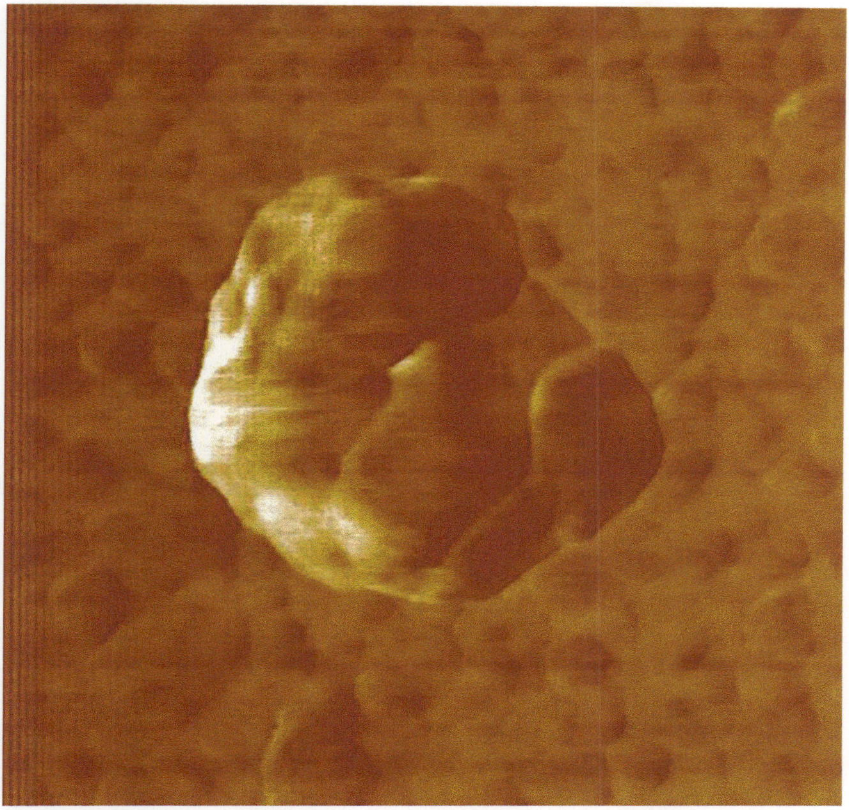

FIGURE 6.3 *Mycobacterium tuberculosis* after 3 months under stress conditions. More than 10–15% of cells under stress conditions show folding phenomena. It seems that rod-shaped *M. tuberculosis* folded from the center in such a way that the two ends of the bacillus (ie, head and tail) approached each other.

entirely missing cell walls (Markova, 2012). More recently, we showed latent *M. tuberculosis* as a complex mixture of acid-fast and nonacid-fast cells. The percentage of cells depends on the time of culture, and the survival strategies are correlated with cell shape, size, and conspicuous loss of acid fastness (Velayati et al., 2011, 2012, 2016; Velayati and Farnia, 2012a,b) (Figs. 6.16–6.20).

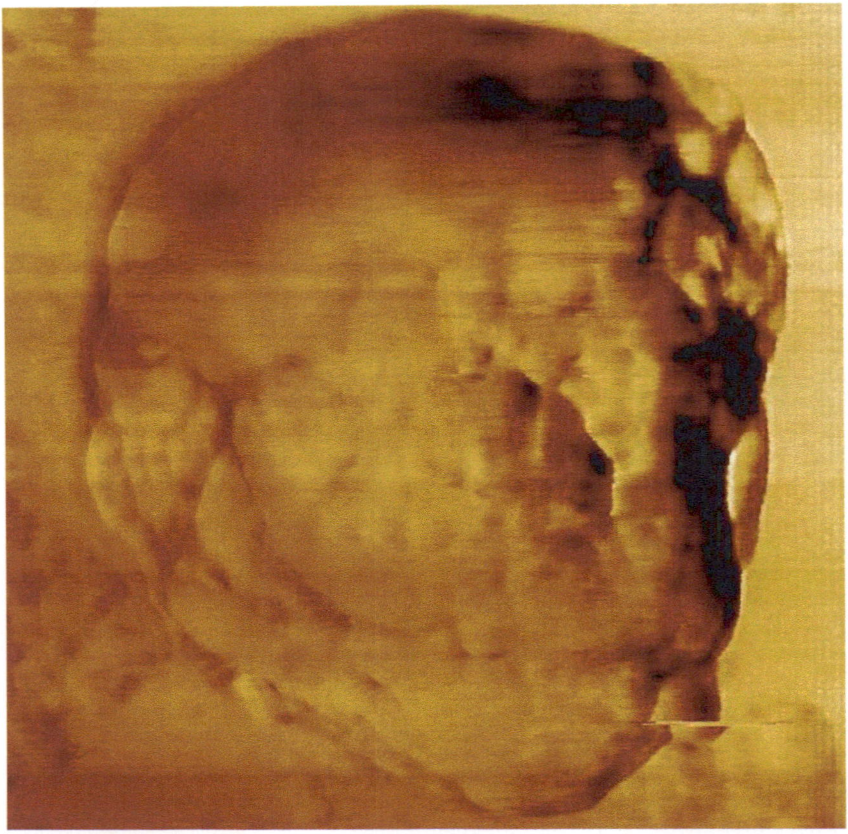

FIGURE 6.4 *Mycobacterium tuberculosis* after 3 months in stress conditions recorded by atomic force microscopy (AFM). In general, such a type of folding cells was reported as round or oval-shaped bacilli by using transmission electron microscopy. Only after using AFM did it become clear that these cells are actually folded *M. tuberculosis*.

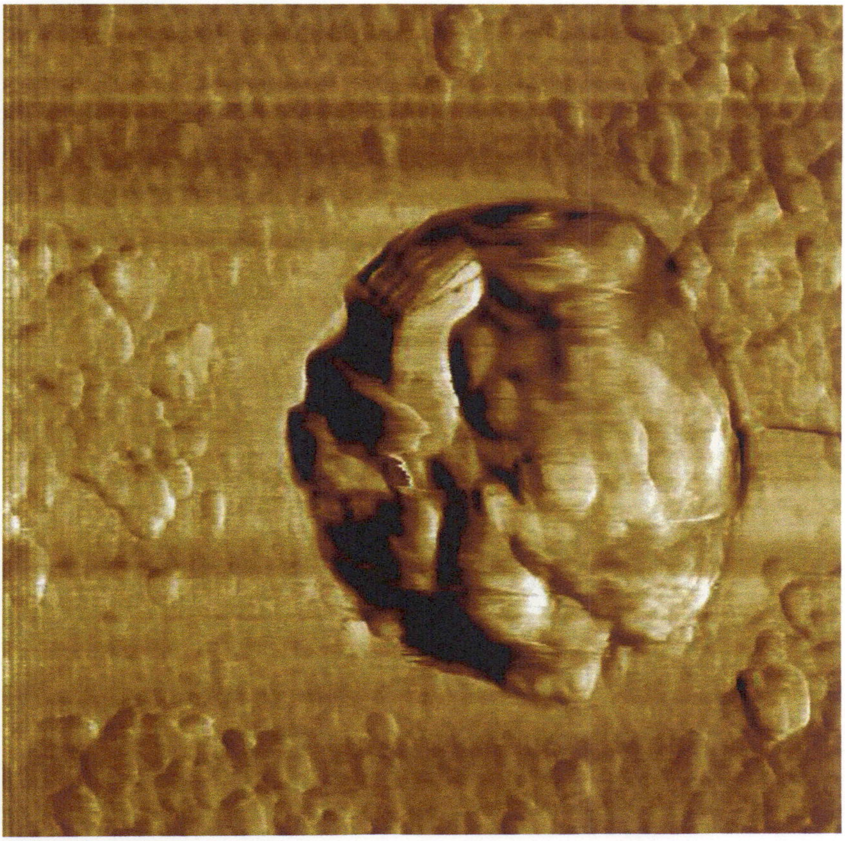

FIGURE 6.5 *Mycobacterium tuberculosis* after 6 months in stress conditions recorded by atomic force microscopy. The number of folded bacilli will increase to 50% of the total population by the end of 6 months of incubation under zero oxygen and nutrients.

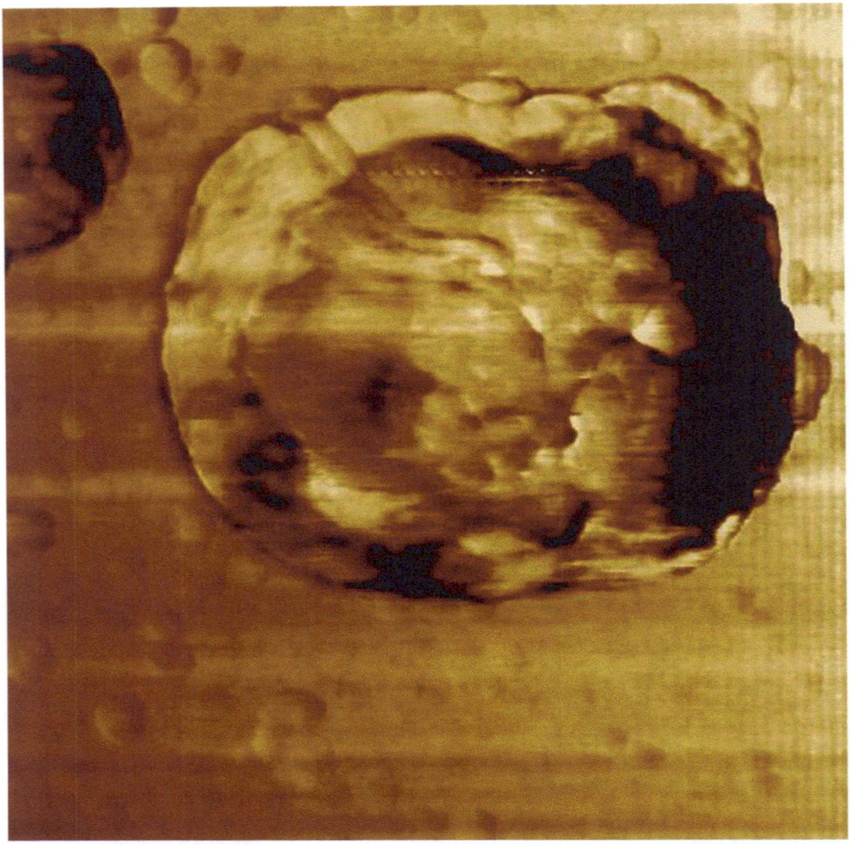

FIGURE 6.6 *Mycobacterium tuberculosis* after 8–10 months in stress conditions recorded by atomic force microscopy. In stress conditions the microorganisms begin to adapt by changing their shape and reducing their size. In this regards, *M. tuberculosis*, similar to other bacteria, depend on laws of diffusion for bringing compounds to their surfaces and mixing them with macromolecules in their cytoplasm. Therefore to import nutrients in harsh environments *M. tuberculosis* must increase their surface area without increasing the surface-to-volume ratio, which can occur when *Mycobacterium* fold and change their shape from a rod to a round shape.

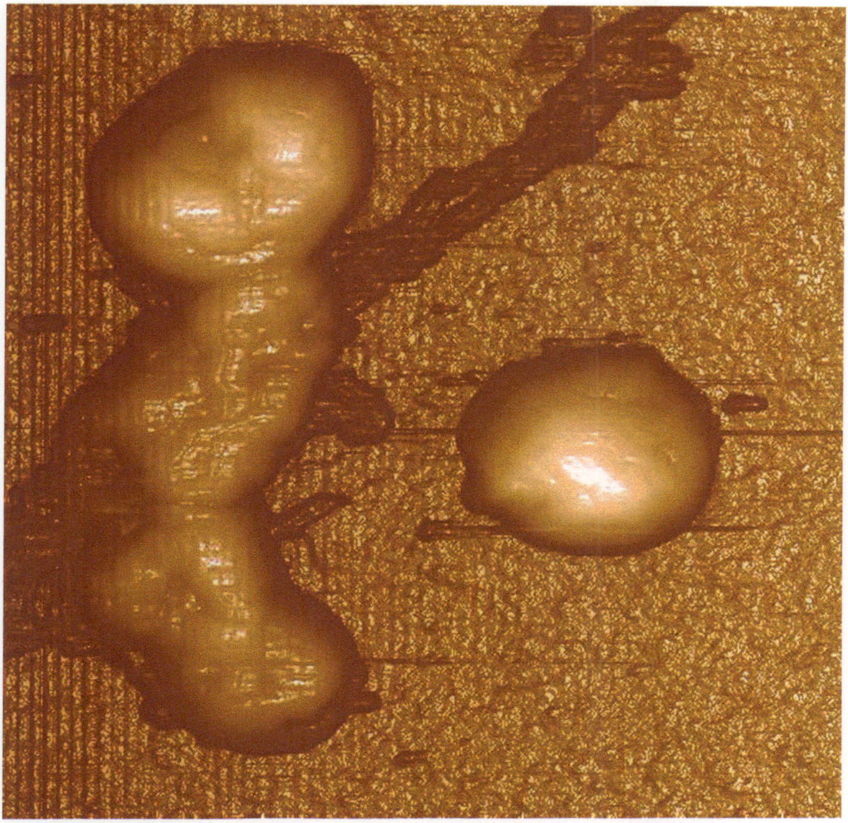

FIGURE 6.7 *Mycobacterium tuberculosis* after 12–18 months in stress conditions recorded by atomic force microscopy. The ovoid- or round-shaped bacilli are usually formed either by folding phenomena or by an asymmetrical or budding type of cell division. Although these cells look very similar in their appearance (by transmission electron microscopy), they have a different surface roughness and cell wall thickening.

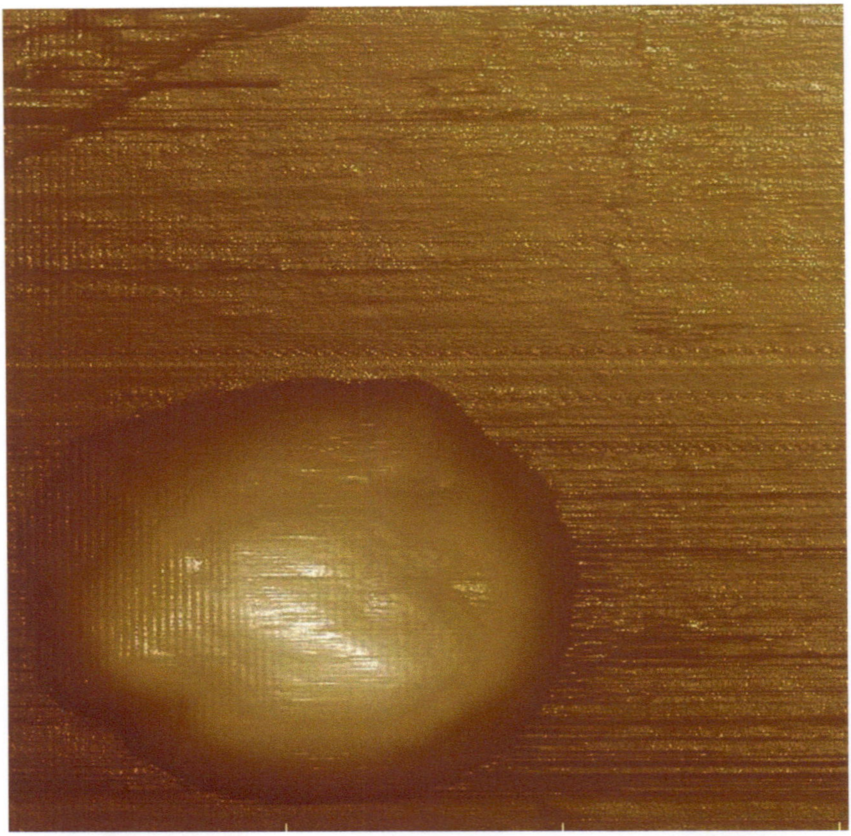

FIGURE 6.8 Ovoid-shaped latent *Mycobacterium tuberculosis* formed by an asymmetrical type of cell division.

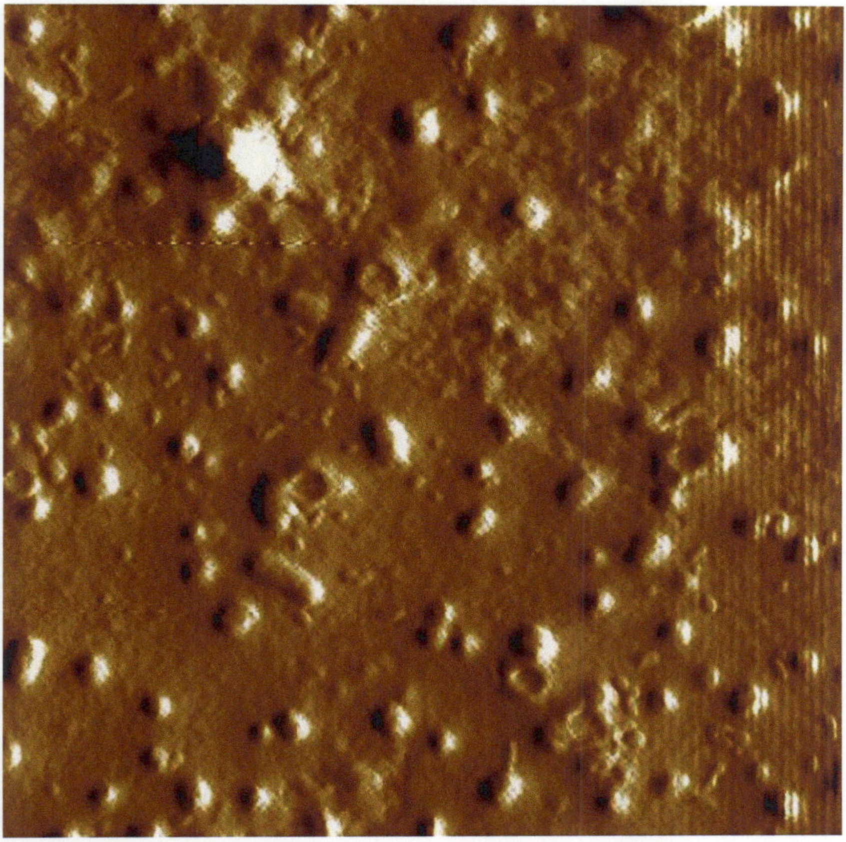

FIGURE 6.9 Culture of *Mycobacterium tuberculosis* after 26 months in stress conditions recorded by the deflection mode of AFM. At the end of 26 months the smears from sediments of culture tubes had a mixed population of rod-shaped (10–15%) or round (85–90%) cells with the average size of 0.5–1 μm.

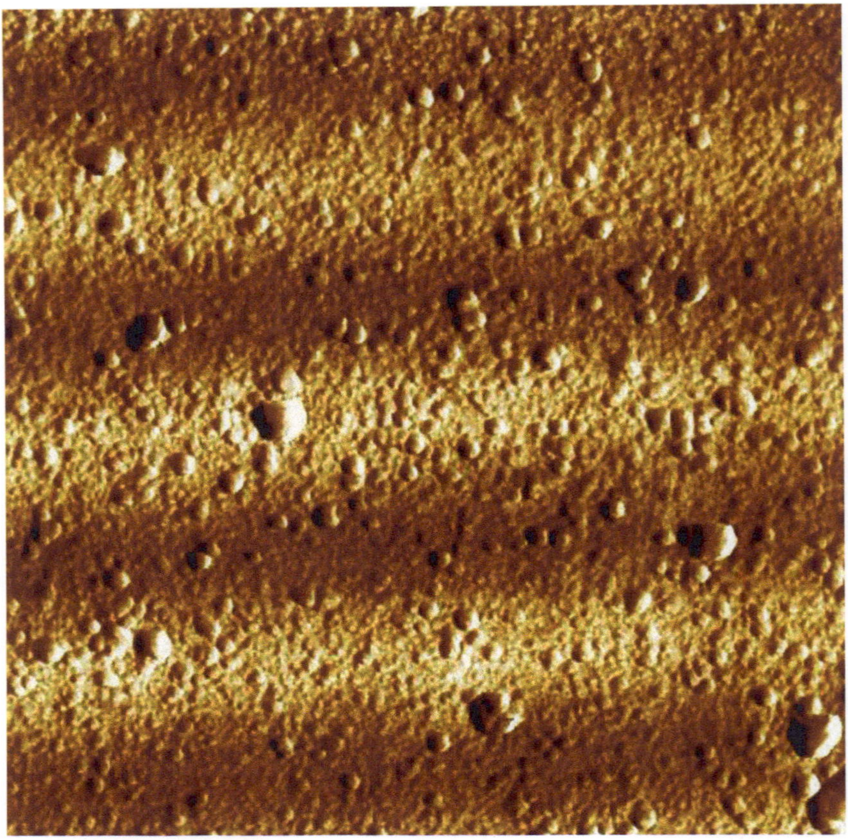

FIGURE 6.10 *Mycobacterium tuberculosis* after 18–26 months under stress conditions recorded by the deflection mode of atomic force microscopy. The cell pellets were obtained by filtering (Millipore filter; 0.2 μm) and centrifugation (14,000 rpm for 30 min) of culture media. The filtered sediments consisted of uniformly nonacid-fast oval- or round-shaped cells with an average size of 0.1–0.3 μm.

FIGURE 6.11 Spore-like cells of *Mycobacterium tuberculosis* detected in anaerobic culture. In the experimental setup the spore was detected after 18 months of dormancy (Velayati et al., 2011). The term *dormancy* is used to describe tuberculosis diseases and the metabolic state of the tubercle bacillus.

FIGURE 6.12 Spore-like cells of *Mycobacterium tuberculosis*. The spore-like cells could not withstand temperatures above 65°C, and they failed to form a heat-resistant colony-forming unit.

FIGURE 6.13 Spore-like cells of *Mycobacterium tuberculosis* recorded by height mode of atomic force microscopy.

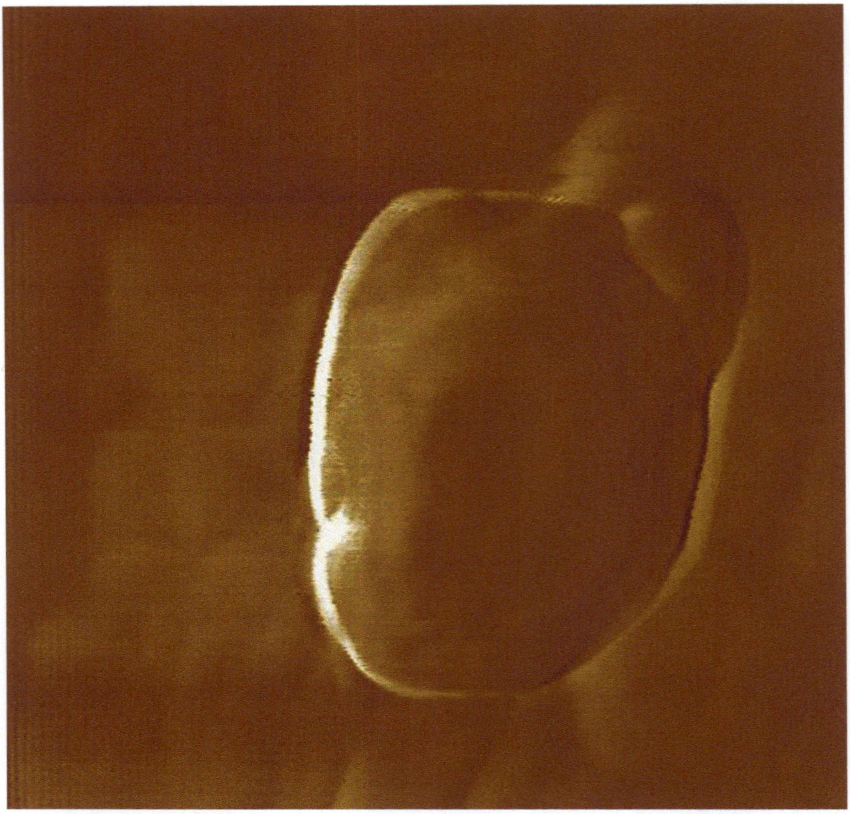

FIGURE 6.14 Rupturing the spore-like cells of *Mycobacterium tuberculosis*. Spore-like cells rupturing and producing small-sized progeny (150–250nm) that are nonacid fast (Velayati et al., 2011).

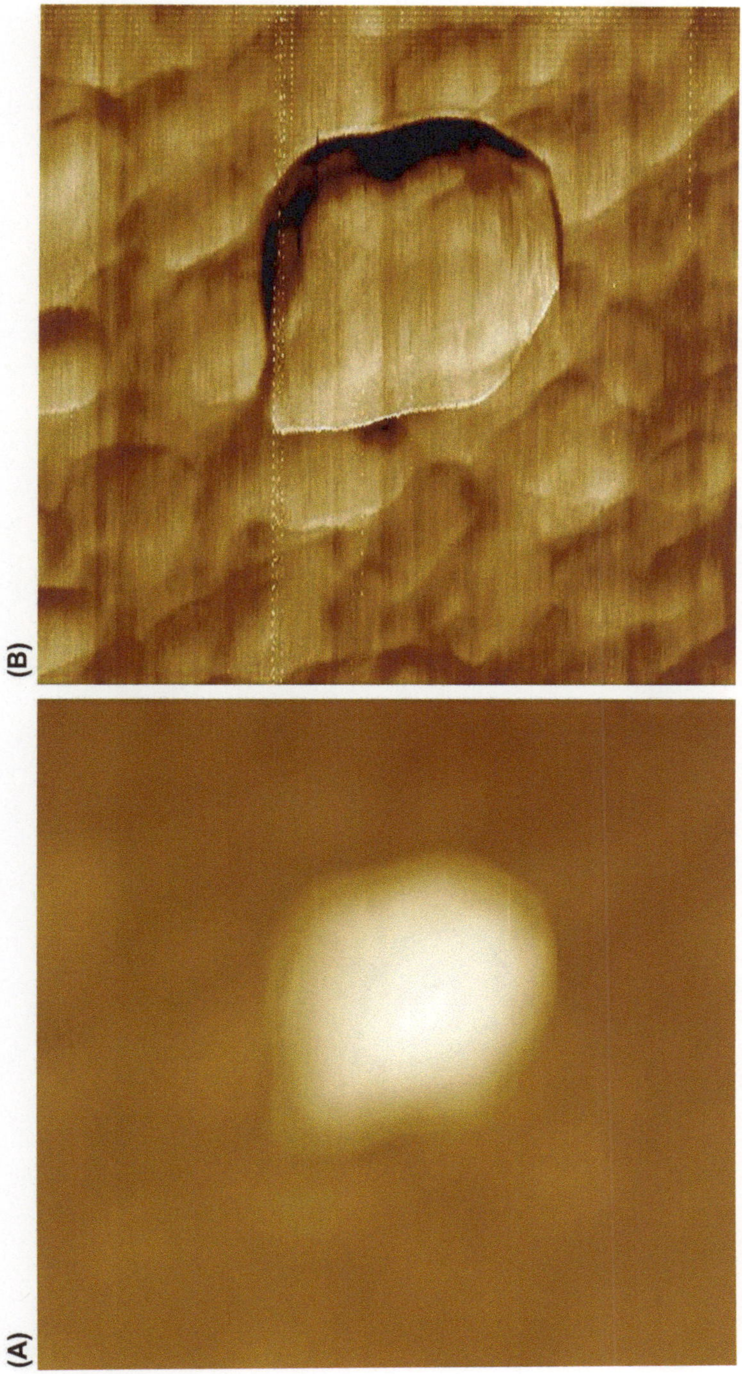

FIGURE 6.15 Nonacid-fast bacilli (cell wall-free) recorded by (A) height and (B) friction mode by atomic force microscopy (Velayati et al., 2016).

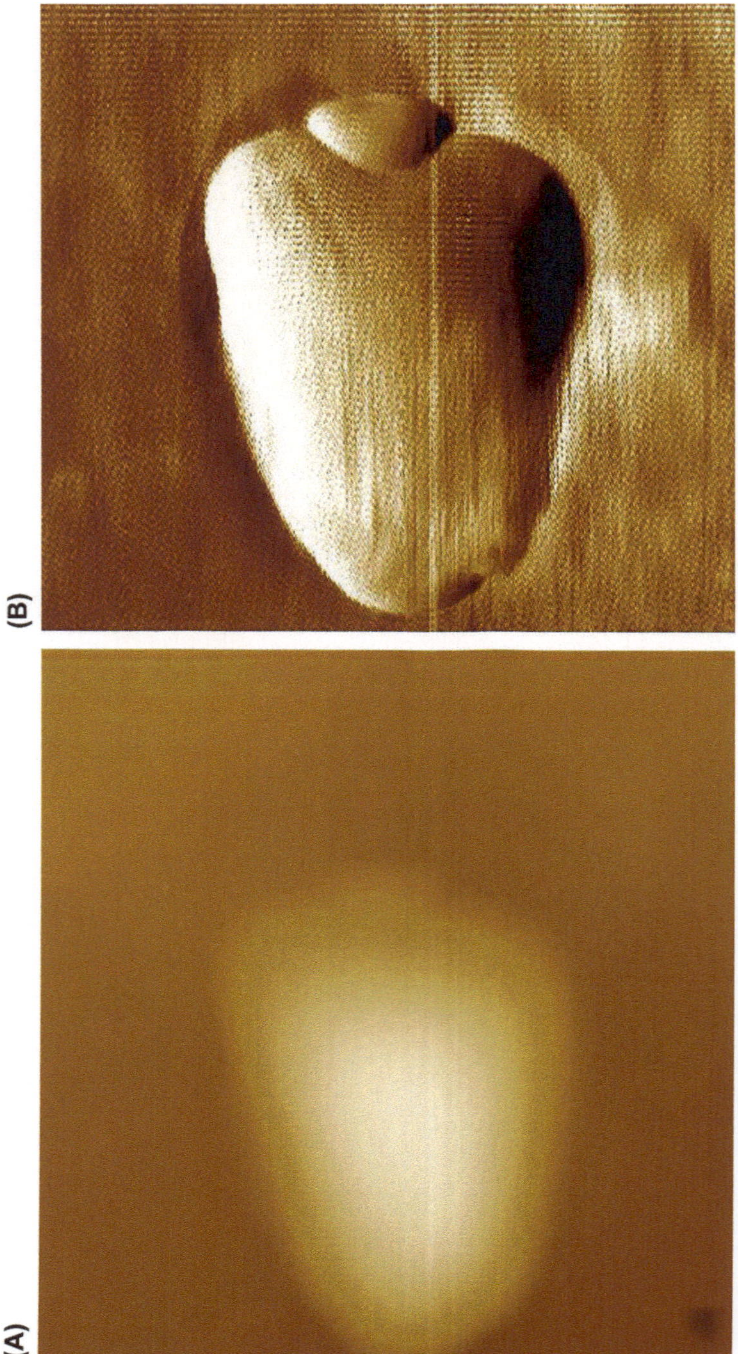

FIGURE 6.16 Persister-like cell detected in blood sample of multidrug-resistant tuberculosis patients after completion of therapy (Velayati et al., 2012) recorded by (A) height and (B) friction mode by atomic force microscopy. As a pathogen *Mycobacterium tuberculosis* manifests its unusual capacity to persist in many ways.

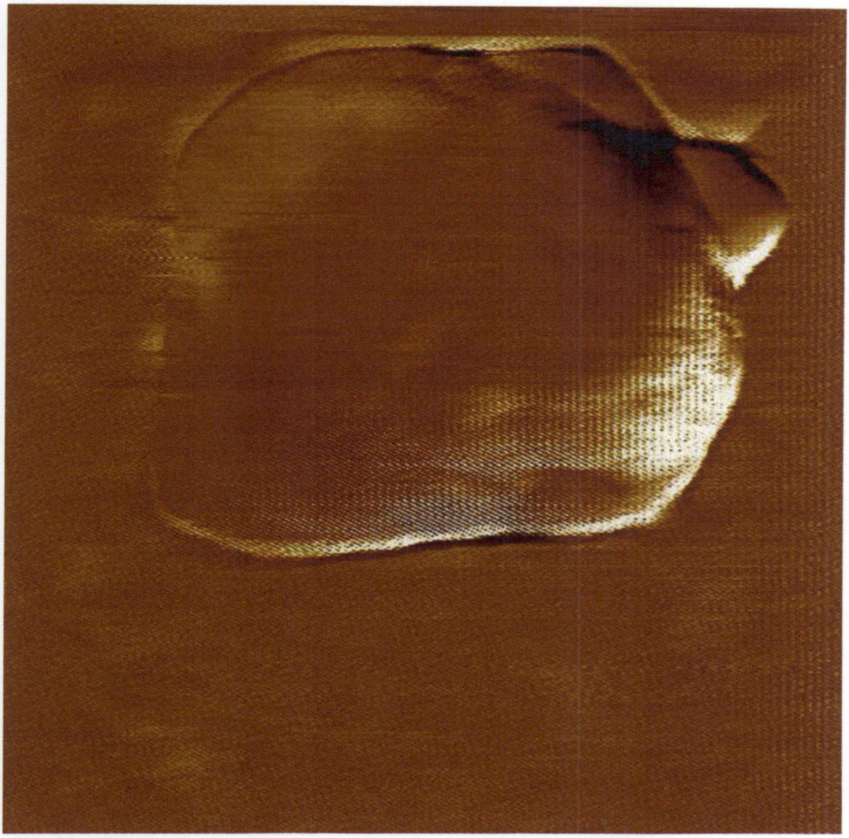

FIGURE 6.17 Persister-like cell of *Mycobacterium tuberculosis* recoded by height mode of atomic force microscopy. To image these cells the silicon surface is charged with polyclonal rabbit anti-mycobacterium antibody. Persister-like cells are small, round or oval in shape, and their numbers are variable in different specimens (Velayati et al., 2012).

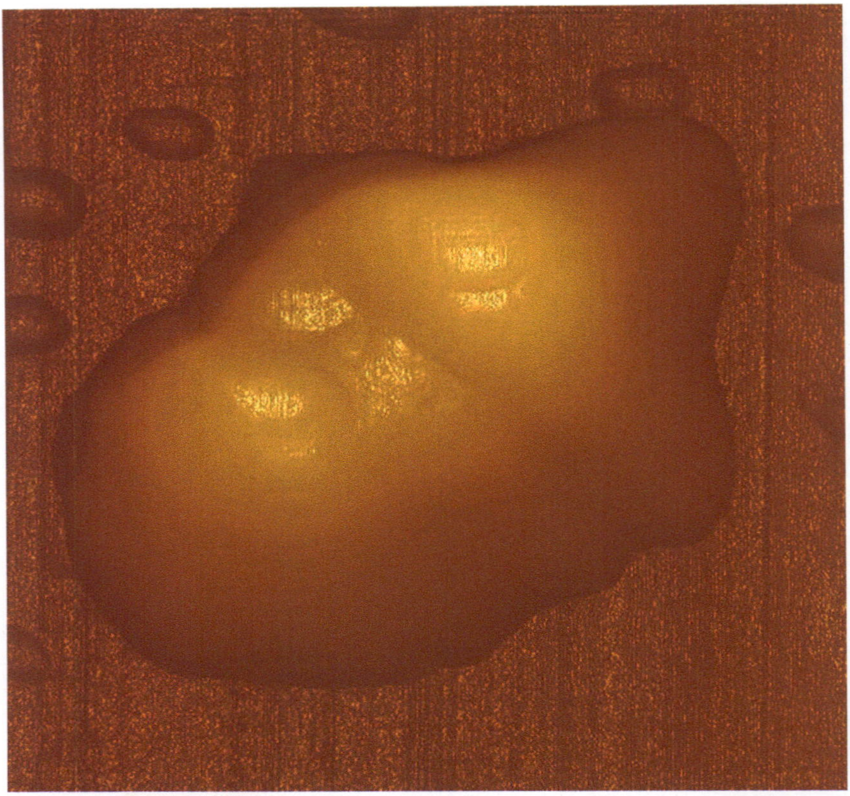

FIGURE 6.18 Dormant cell of *Mycobacterium tuberculosis* preparing for division cycle recorded by height mode under atomic force microscopy. Dormancy is strongly associated with an in vitro model of *M. tuberculosis* growth under limiting oxygen tension. Whether these dormant cells are capable of dividing is not yet clear and requires further investigation.

FIGURE 6.19 Cell-wall–defiant *Mycobacterium tuberculosis* preparing for division cycle recoded by height mode under atomic force microscopy.

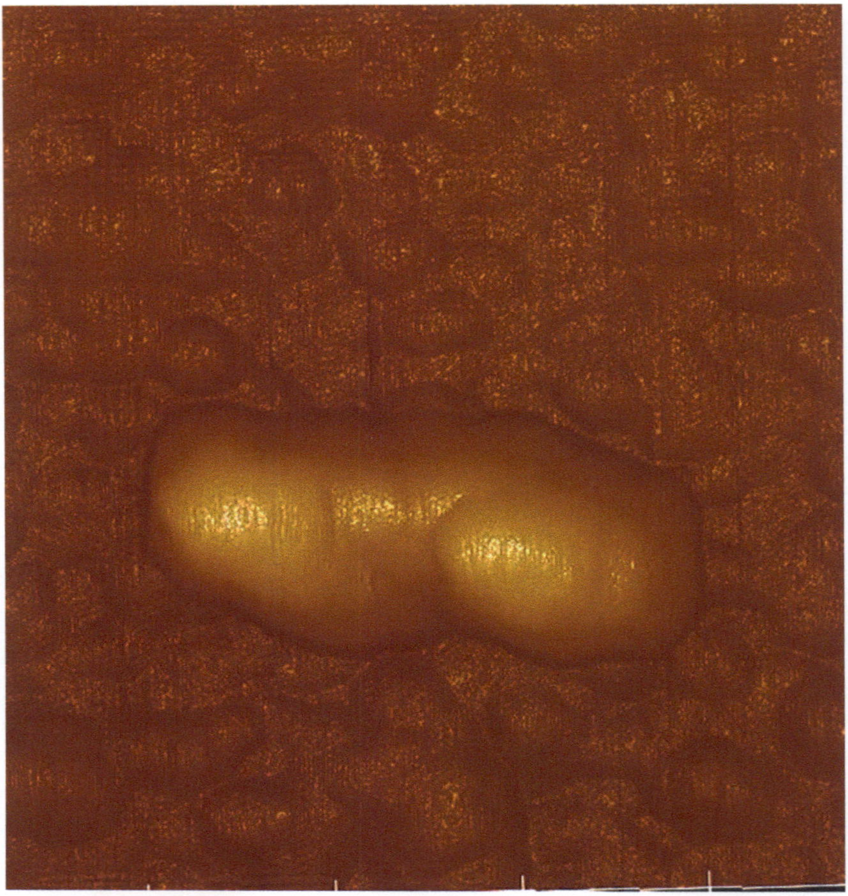

FIGURE 6.20 Rod-shaped latent *Mycobacterium tuberculosis* recorded by height mode under atomic force microscopy. Ten to fifteen percent of the total population under dormant condition remained acid fast and had an original shape with an average size of 0.2–0.5 µm.

References

Acosta, C.D., Dadu, A., Ramsay, A., et al., 2014. Drug-resistant tuberculosis in Eastern Europe: challenges and ways forward. Public Health Action 4, S3–S12.

Adam, A., Petit, J.F., Wietzerbin-Falszpan, J., et al., 1969. L'acide N-glycolyl-muramique, constituant des parois de *Mycobacterium smegmatis*; identification par spectromttrie de masse. FEBS Lett. 4, 87–92.

Akamatsu, Y., Ono, Y., Nojima, S., 1966. Phospholipid patterns in subcellular fractions of *Mycobacterium phlei*. J. Biochem. 59, 176–182.

Aldridge, B.B., Fernandez-Suarez, M., Heller, D., et al., 2012. Asymmetry and aging of mycobacterial cells lead to variable growth and antibiotic susceptibility. Science 335, 100–104.

Alexander, K.A., Laver, P.N., Michel, A.L., et al., 2010. Novel *Mycobacterium tuberculosis* complex pathogen, *M. mungi*. Emerg. Infect. Dis. 16, 1296–1299.

Alland, D., Kramnik, I., Weisbrod, T.R., et al., 1998. Identification of differentially expressed mRNA in prokaryotic organisms by customized amplification libraries (DECAL): the effect of isoniazid on gene expression in *Mycobacterium tuberculosis*. Proc. Natl. Acad. Sci. U.S.A. 95, 13227–13232.

Alsteens, D., Verbelen, C., Dague, E., et al., 2008. Organization of the mycobacterial cell wall: a nanoscale view. Pflugers Arch. 456, 117–125.

Alteri, C.J., Xicohténcatl-Cortes, J., Hess, S., et al., 2007. *Mycobacterium tuberculosis* produces pili during human infection. Proc. Natl. Acad. Sci. U.S.A. 104, 5145–5150.

Amberson, J.B., 1938. The significance of latent forms of tuberculosis. N. Engl. J. Med. 219, 572–576.

American Trudeau Society, 1953. Report of a panel discussion on survival, revival of tubercle bacilli in healed and tuberculous lesions. Am. Rev. Tuberc. 68, 477–495.

Andersen, K., 1948. A report on BCG vaccinations in Copenhagen. J. Am. Med. Womens Assoc. 3, 399–401.

Andersen, P., Urdahl, K.B., 2015. TB vaccines; promoting rapid and durable protection in the lung. Curr. Opin. Immunol. 35, 55–62.

Andersen, P., Askgaard, D., Ljungqvist, L., et al., 1991. Proteins released from *Mycobacterium tuberculosis* during growth. Infect. Immun. 59, 1905–1910.

Anuchin, A.M., Mulyukin, A.L., Suzina, N.E., et al., 2009. Dormant forms of *Mycobacterium smegmatis* with distinct morphology. Microbiology 155, 1071–1079.

Ariji, F., Yamaguchi, J., Fukushi, K., Oka, S., 1968. Electron microscopic localization of enzymes in *Mycobacterium tuberculosis*. Sci. Rep. Res. Inst. Tohoku Univ. Med. 15, 88–97.

Arima, J., Takahashi, S., Igarashi, H., 1972. Optic and electron microscope study of cross-bands of photochromogenic mycobacteria. C.R. Seances Soc. Biol. Fil. 166, 480–483.

Aronson, J.D., 1948. The evaluation of BCG vaccine in the control of tuberculosis. Trans. Stud. Coll. Physicians Phila 16, 1–13.

Asselineau, J., Lederer, E., 1950. Structure of the mycolic acids of mycobacteria. Nature 166, 782–783.

Baess, I., 1974. Isolation and purification of deoxyribonucleic acid from mycobacteria. Acta Pathol. Microbiol. Scand. B Microbiol. Immunol. 82, 780–784.

Baker, L., Brown, T., Maiden, M.C., et al., 2004. Silent nucleotide polymorphisms and a phylogeny for *Mycobacterium tuberculosis*. Emerg. Infect. Dis. 10, 1568–1577.

Balasubramanian, V., Wiegeshaus, E.H., Taylor, B.T., et al., 1994. Pathogenesis of tuberculosis: pathway to apical localization. Tuber. Lung Dis. 75, 168–178.

Banerjee, A., Dubnau, E., Ouernard, A., et al., 1994. inhA, a gene encoding a target for isoniazid and ethionamide in *Mycobacterium tuberculosis*. Science 263, 227–230.

Barák, I., Wilkinson, A.J., 2007. Division site recognition in *Escherichia coli* and *Bacillus subtilis*. FEMS Microbiol. Rev. 31, 311–326.

Barksdale, L., Kim, K.S., 1977. Mycobacterium. Bacteriol. Rev. 41, 217–372.

Barnes, P.F., Bloch, A.B., Davidson, P.T., et al., 1991. Tuberculosis in patients with human immunodeficiency virus infection. N. Engl. J. Med. 324, 1644–1650.

Barrow, W.W., Ullom, B.P., Brennan, P.J., 1980. Peptidoglycolipid nature of the superficial cell wall sheath of smooth-colony-forming mycobacteria. J. Bacteriol. 144, 814–822.

Barry 3rd, C.E., 2001. Interpreting cell wall 'virulence factors' of *Mycobacterium tuberculosis*. Trends Microbiol. 9, 237–241.

Barry 3rd, C.E., Boshoff, H.I., Dartois, V., 2009. The spectrum of latent tuberculosis: rethinking the biology and intervention strategies. Nat. Rev. Microbiol. 7, 845–855.

Bates, J.H., Stead, W.W., 1993. The history of tuberculosis as a global epidemic. Med. Clin. North Am. 77, 1205–1217.

Beck, F., Yegian, D., 1952. A study of the tubercle bacillus in resected pulmonary lesions. Am. Rev. Tuberc. 66, 44–50.

Benedek, T.G., 2004. The history of gold therapy for tuberculosis. J. Hist. Med. Allied Sci. 59, 50–89.

Benedetti, E.L., Dunia, I., Ludosky, M.A., et al., 1984. Freeze-etching and freeze-fracture structural features of cell envelopes in mycobacteria and leprosy derived corynebacteria. Acta Leprol. 2, 237–248.

Betts, J.C., Lukey, P.T., Robb, L.C., et al., 2002. Evaluation of a nutrient starvation model of *Mycobacterium tuberculosis* persistence by gene and protein expression profiling. Mol. Microbiol 43, 717–731.

Bi, E.F., Lutkenhaus, J., 1991. FtsZ ring structure associated with division in *Escherichia coli*. Nature 354, 161–164.

Böddinghaus, B., Rogall, T., Flohr, T., et al., 1990. Detection and identification of mycobacteria by amplification of rRNA. J. Clin. Microbiol. 28, 1751–1759.

Bojalil, L.F., Cerbon, J., Trujillo, A., 1962. Adansonian classification of mycobacteria. J. Gen. Microbiol. 28, 333–346.

Bongiorno, V., 1953. Morphological and chemical changes of *Mycobacterium tuberculosis* after antibiotic therapy. Riv. Ital. Ig. 13, 356–361.

Borgdorff, M.W., Nagelkerke, N.J., Dye, C., et al., 2000. Gender and tuberculosis: a comparison of prevalence surveys with notification data to explore sex differences in case detection. Int. J. Tuberc. Lung Dis. 4, 123–132.

Bradley, S.G., 1973. Relationships among mycobacteria and nocardiae based upon deoxyribonucleic acid reassociation. J. Bacteriol. 113, 645–651.

Bradley, S.G., Bond, J.S., 1974. Taxonomic criteria for Mycobacteria and nocardiae. Adv. Appl. Microbiol. 18, 131–190.

Brennan, P.J., 1994. The microbiology of *Mycobacterium leprae*, part II. Reflections on major developments and those responsible for them. Int. J. Lepr. Other Mycobact. Dis. 62, 594–598.

Brennan, P.J., Nikaido, H., 1995. The envelope of mycobacteria. Annu. Rev. Biochem. 64, 29–63.

Brieger, E.M., Fell, B., 1945. Warm-stage observation on the initial development of the avian tubercle bacillus cultivated in embryo extract. J. Hyg. Camb. 44, 158–165.

Brieger, E.M., Glauert, A.M., 1952. A phase-contrast study of reproduction in mycelia strains of avian tubercle bacilli. J. Gen. Microbiol. 7, 287–302.

Brodie, A.F., Lee, S.H., Kalra, V.K., 1979. Transport and energy transduction mechanism in *Mycobacterium phlei*. In: Schlessinger, D. (Ed.), Microbiology. ASM Press, Washington, D.C., pp. 46–53.

Brosch, R., Gordon, S.V., Marmiesse, M., et al., 2002. A new evolutionary scenario for the *Mycobacterium tuberculosis* complex. Proc. Natl. Acad. Sci. U.S.A. 99, 3684–3689.

Brudey, K., Driscoll, J.R., Rigouts, L., et al., 2006. *Mycobacterium tuberculosis* complex genetic diversity: mining the fourth international spoligotyping database (SpolDB4) for classification, population genetics and epidemiology. BMC Microbiol. 6, 23.

Burdett, I.D., Higgins, M.L., 1978. Study of pole assembly in *Bacillus subtilis* by computer reconstruction of septal growth zones seen in central, longitudinal thin sections of cells. J. Bacteriol. 133, 959–971.

Cameron, T.A., Zupan, J.R., Zambryski, P.C., 2015. The essential features and modes of bacterial polar growth. Trends Microbiol. 23, 347–353.

Caminero, J.A., Sotgiu, G., Zumla, A., et al., 2010. Best drug treatment for multidrug-resistant and extensively drug-resistant tuberculosis. Lancet Infect. Dis. 10, 621–629.

Caminero Luna, J.A., 2015. Update on the diagnosis and treatment of pulmonary tuberculosis. Rev. Clin. Esp. (15), 00230–00231 pii:S0014-2565.

Cardona, P.J., 2007. New insights on the nature of latent tuberculosis infection and its treatment. Inflamm. Allergy Drug Targets 6, 27–39.

Cardona, P.J., Ruiz-Manzano, J., 2004. On the nature of *Mycobacterium tuberculosis* – latent bacilli. Eur. Respir. J. 24, 1044–1051.

Castelnuovo, G., Bellezza, G., Giuliani, H.I., et al., 1968. Chemical and immunologic relationships in the Actinomycetales. II. Serologic relationships between the streptomycetes, nocardiae and mycobacteria. Ann. Inst. Pasteur (Paris) 114, 139–147.

Centers for Disease Control and Prevention (CDC), 1990. Nosocomial transmission of multidrug-resistant tuberculosis to health-care workers and HIV infected patients in an urban hospital – Florida. MMWR Morb. Mortal. Wkly. Rep. 39, 718–722.

Centers for Disease Control and Prevention (CDC), 1991. Nosocomial transmission of multidrug-resistant tuberculosis among HIV-infected persons – Florida and New York, 1988–1991. MMWR Morb. Mortal. Wkly. Rep. 40, 585–591.

Centers for Disease Control and Prevention (CDC), 2006. Emergence of *Mycobacterium tuberculosis* with extensive resistance to second-line drugs- worldwide, 2000–2004. MMWR Morb. Mortal. Wkly. Rep. 55, 301–305.

Chandrasekhar, S., Ratnam, S., 1992. Studies on cell-wall deficient non-acid fast variants of *Mycobacterium tuberculosis*. Tuber. Lung Dis. 73, 273–279.

Chao, M.C., Rubin, E.J., 2010. Letting sleeping dos lie: does dormancy play a role in tuberculosis? Annu. Rev. Microbiol. 64, 293–311.

Chauhan, A., Lofton, H., Maloney, E., et al., 2006. Interference of *Mycobacterium tuberculosis* cell division by Rv2719c, a cell wall hydrolase. Mol. Microbiol. 62, 132–147.

Chen, J.C., Beckwith, J., 2001. FtsQ, FtsL and FtsI require FtsK, but not FtsN, for co-localization with FtsZ during *Escherichia coli* cell division. Mol. Microbiol. 42, 395–413.

Cohen, J., 2006. Infectious disease. Extensively drug-resistant TB gets foothold in South Africa. Science 313, 1554.

Colijn, C., Cohen, T., Ganesh, A., et al., 2011. Spontaneous emergence of multiple drug resistance in tuberculosis before and during therapy. PLoS One 8, e18327.

Collins, C.H., Grange, J.M., Yates, M.D., 1984. Mycobacteria in water. J. Appl. Bacteriol. 57, 193–211.

Collins, C.H., Yates, M.D., Uttley, A.H., 1988. Presumptive identification of nocardias in a clinical laboratory. J. Appl. Bacteriol. 65, 55–59.

Cook, G.M., Berney, M., Gebhard, S., et al., 2009. Physiology of mycobacteria. Adv. Microb. Physiol. 55, 81–319.

Corper, H.J., Cohn, M.L., 1933. The viability and virulence of old cultures of tubercle bacilli: studies on twelve-year broth cultures maintained at incubator temperature. Am. Rev. Tuberc. 28, 856–874.

Crofton, J., 1959. Chemotherapy of pulmonary tuberculosis. BMJ 1, 1610–1614.

Crofton, J., 1970. The assessment and treatment of drug resistance problems in tuberculosis. J. Ir. Med. Assoc. 63, 75–78.

Cross, T., Goodfellow, M., 1973. In: Sykes, G., Skinner, F.A. (Eds.), Actinomycetabs: Characteristics and Practical Importance. Academic Press, New York, pp. 11–112.

Csillage, A., 1962. Development of form 2 mycobacterium on autoclaved LJ medium. Tubercle 43, 433–439.

Csillage, A., 1964. The mycococcus form of mycobacteria. J. Gen. Microbiol. 34, 341–352.

Csillage, A., 1970. A simple method to obtain the mycococcus form of *Mycobacterium phlei*. J. Gen. Microbiol. 62, 251–259.

Cunningham, A.F., Spreadbury, C.L., 1998. Mycobacterial stationary phase induced by low oxygen tension: cell wall thickening and localization of the 16-kilodalton α –crystallin homolog. J. Bacteriol. 180, 801–808.

Cunto, G., Kanetsuna, F., Imaeda, T., 1969. Chemical analysis of the mucopeptide of *Mycobacterium smegmatis*. Biochim. Biophys. Acta 192, 358–360.

Da Silva, P.E.A., Palomino, J.C., 2011. Molecular basis and mechanisms of drug resistance in *Mycobacterium tuberculosis* classical and new drugs. J. Antimicrob. Chemother. 66, 1417–1430.

Daffé, M., Draper, P., 1998. The envelope layers of mycobacteria with reference to their pathogenicity. Adv. Microb. Physiol. 39, 131–203.

Daffe, M., Etienne, G., 1999. The capsule of *Mycobacterium tuberculosis* and its implications for pathogenicity. Tuber. Lung Dis. 79, 153–169.

Daffé, M., Reyrat, J.M., 2008. The Mycobacterial Cell Envelope, first ed. ASM Press, USA.

Daffe, M., Lankelle, M.A., Puzo, G., 1983. Structural elucidation by field desorption and electron-impact mass spectrometry of the C-mycosides isolated from *Mycobacterium smegmatis*. Biochim. Biophys. Acta 751, 439–443.

Dahl, J.L., 2004. Electron microscopy analysis of *Mycobacterium tuberculosis* cell division. FEMS Microbiol. Lett. 240, 15–20.

Dajkovic, A., Lutkenhaus, J., 2006. Z ring as executor of bacterial cell division. J. Mol. Microbiol. Biotechnol. 11, 140–151.

D'Ambrosio, L., Tadolini, M., Centis, R., et al., 2015. Supporting clinical management of the difficult-to-treat TB cases: the ERS-WHO TB Consilium. Int. J. Infect. Dis. 32, 156–160.

Daniel, T.M., 2006. The history of tuberculosis. Respir. Med. 100, 1862–1870.

Danilchanka, O., Sun, J., Pavlenok, M., et al., 2014. An outer membrane channel protein of *Mycobacterium tuberculosis* with exotoxin activity. Proc. Natl. Acad. Sci. U.S.A. 111, 6750–6755.

Dannenberg Jr., A.M., 1993. Immunopathogenesis of pulmonary tuberculosis. Hosp. Pract. (Off. Ed.) 28, 51–58.

David, H.L., 1970. Probability distribution of drug resistant mutants in unselected populations of *Mycobacterium tuberculosis*. Appl. Microbiol. 20, 810–814.

David, H.L., 1974. Carotenoid pigments of *Mycobacterium kansasii*. Appl. Microbiol. 28, 696–699.

Davis, J., 1994. Inactivation of antibiotics and the dissemination of resistance genes. Science 264, 375–382.

Deb, C., Lee, C.M., Dubey, V., et al., 2009. A novel in vitro multiple-stress dormancy model for *Mycobacterium tuberculosis* generates a lipid-loaded, drug-tolerant, dormant pathogen. PLoS One 4, e6077.

Delogu, G., Provvedi, R., Sali, M., et al., 2015. *Mycobacterium tuberculosis* virulence: insights and impact on vaccine development. Future Microbiol. 10, 1177–1194.

Donoghue, H.D., Spigelman, M., Greenblatt, C.L., et al., 2004. Tuberculosis: from prehistory to Robert Koch, as revealed by ancient DNA. Lancet Infect. Dis. 4, 584–592.

Donovan, C., Bramkamp, M., 2014. Cell division in Corynebacterineae. Front. Microbiol. 5, 132–148.

Draper, P., 1982. Bacteriology of *Mycobacterium leprae*: state of the art paper. Ann. Microbiol. (Paris) 133, 13–14.

Draper, P., 1998. The outer parts of the mycobacterial envelope as permeability barriers. Front. Biosci. 3, 1253–1261.

Draper, P., Daffé, M., 2005. The cell envelope of *Mycobacterium tuberculosis* with special reference to the capsule and outer permeability barrier. In: Cole, S., Eisenach, K., McMurray, D., et al. (Eds.), Tuberculosis and the Tubercle Bacillus. ASM Press, Washington, D.C., pp. 261–274.

Draper, P., Rees, R.J.W., 1970. Electron-transparent zone of mycobacteria may be a defence mechanism. Nature 228, 860–861.

Dye, C., 2006. Global epidemiology of tuberculosis. Lancet 367, 938–940.

Dziadek, J., Rutherford, S.A., Madiraju, M.V., et al., 2003. Rajagopalan M Conditional expression of Mycobacterium smegmatis ftsZ, an essential cell division gene. Microbiology 149, 1593–1603.

Edlin, B.R., Tokars, J.I., Grieco, M.H., et al., 1992. An outbreak of tuberculosis caused by multiple-drug resistant tuberculosis among hospitalized patients with the acquired immunodeficiency syndrome. N. Engl. J. Med. 326, 1514–1521.

Edson, N.L., 1951. The intermediary metabolism of the mycobacteria. Bacteriol. Rev. 15, 147–182.

Edwards, R.P., 1970. Electron-microscope illustrations of division in *Mycobacterium leprae*. J. Med. Microbiol. 3, 493–499.

Ehrlich, P., 1887. Uber die Methylenblaureaktion der lebenden Nervensubstanz. Biol. Zbl. 6, 214–224.

Eldholm, V., Matee, M., Mfinanga, S.G., et al., 2006. A first insight into the genetic diversity of *Mycobacterium tuberculosis* in Dar es Salaam, Tanzania, assessed by spoligotyping. BMC Microbiol. 6, 76.

Embley, T.M., Stackebrandt, E., 1994. The molecular phylogeny and systematics of the actinomycetes. Annu. Rev. Microbiol. 48, 257–289.

Esmail, H., Barry 3rd, C.E., Young, D.B., et al., 2014. The ongoing challenge of latent tuberculosis. Phil. Trans. R. Soc. B 369, 20130437.

Etemadi, A.H., 1967. Codlations structurales et biogtnttiques des acides mycoliques en rapport avec la phylogenbse de quelques genres d'actinomycktales. Bull. Soc. Chim. Biol. 49, 695–706.

Etemadi, A.H., Pinte, F., Markovits, J., 1966. Sur la structure des prkurseurs Cventuels des acides u-smegmamycoliques. C.R. Acad. Sci. Paris 262, 1343–1346.

Falkinham III, J.O., 1996. Epidemiology of infection by nontuberculous mycobacteria. Clin. Microbiol. Rev. 9, 177–215.

Falkinham III, J.O., 2002. Nontuberculous mycobacteria in the environment. Clin. Chest Med. 23, 529–551.

Falkinham III, J.O., 2009. Surrounded by mycobacteria: nontuberculous mycobacteria in the human environment. J. Appl. Microbiol. 107, 356–367.

Falkinham III, J.O., 2010. Impact of human activities on the ecology of nontuberculous mycobacteria. Future Microbiol. 5, 951–960.

Falzon, D., Weyer, K., Raviglione, M.C., 2013. Drug-resistant tuberculosis: latest advances. Lancent Respir. Med. 1, e9–e10.

Farnia, P., Masjedi, M.R., Farnia, P., et al., 2010. Growth and cell-division in extensive (XDR) and extremely drug resistany (XXDR) tuberculosis strains: transmission and atomic force observation. Int. J. Clin. Exp. Med. 3, 308–314.

Feldman, W.H., 1938. The residual infectivity of the primary complex of tuberculosis. Am. J. Pathol. 14, 473–490.

Feldman, W.H., Baggenstoss, A.H., 1939. The occurrence of virulent tubercle bacilli in presumably non-tuberculous lung tissue. Am. J. Pathol. 15, 501–515.

Filliol, I., Driscoll, J.R., Van Soolingen, D., et al., 2002. Global distribution of *Mycobacterium tuberculosis* spoligotypes. Emerg. Infect. Dis. 8, 1347–1349.

Filliol, I., Motiwala, A.S., Cavatore, M., et al., 2006. Global phylogeny of *Mycobacterium tuberculosis* based on single nucleotide polymorphism (SNP) analysis: insights into tuberculosis evolution, phylogenetic accuracy of other DNA fingerprinting systems, and recommendations for a minimal standard SNP set. J. Bacteriol. 188, 759–772.

Fischel, F., 1893. Zur Morphologie und Biologiedes Tuberkelbacillus.Klin. Wochschr 30, 989–993.

Fischl, M.A., Uttamchandani, R.B., Daikos, G.L., et al., 1992. An outbreak of tuberculosis caused by multiple-drug resistant tubercle bacilli among patients with HIV infection. Ann. Intern. Med. 117, 177–183.

Flynn, J.L., Chan, J., 2001. Tuberculosis: latency and reactivation. Infect. Immun. 69, 4195–4201.

Fomukong, N.G., Tang, T.H., al-Maamary, S., et al., 1994. Insertion sequence typing of *Mycobacterium tuberculosis*: characterization of a widespread subtype with a single copy of IS6110. Tuber. Lung Dis. 75, 435–440.

Forrellad, M.A., Klepp, L.I., Gioffré, A., et al., 2013. Virulence factors of the *Mycobacterium tuberculosis* complex. Virulence 4, 3–66.

Fox, W., Sutherland, I., Daniels, M., 1954. A five-year assessment of patients in a controlled trial of streptomycin in pulmonary tuberculosis. Q.J. Med. 23, 347–366.

Fox, W., Ellard, G.A., Mitchison, D.A., 1999. Studies on the treatment of tuberculosis undertaken by the British Medical Research Council tuberculosis units, 1946–1986, with relevant subsequent publications. Int. J. Tuberc. Lung Dis. 3, S231–S279.

Fregnan, G.B., Smith, D.W., 1962. Description of various colony forms of mycobacteria. J. Bacteriol. 83, 819–827.

Frieden, T.R., Sterling, T., Pablos-Mendez, A., et al., 1993. The emergence of drug resistance tuberculosis in New York City. N. Engl. J. Med. 328, 521–526.

Frothingham, R., Meeker-O'Connell, W.A., 1998. Genetic diversity in the *Mycobacterium tuberculosis* complex based on variable numbers of tandem DNA repeats. Microbiology 144, 1189–1196.

García-Rodríguez, J.F., Álvarez-Díaz, H., Lorenzo-García, M.V., et al., 2011. Extrapulmonary tuberculosis: epidemiology and risk factors. Enferm. Infecc. Microbiol. Clin 29, 502–509.

Garton, N.J., Christensen, H., Minnikin, D.E., et al., 2002. Intracellular lipophilic inclusions of mycobacteria in vitro and in sputum. Microbiology 148, 2951–2958.

Gengenbacher, M., Rao, S.P.S., Pethe, K., et al., 2010. Nutrient-starved, non-replicating *Mycobacterium tuberculosis* requires respirations, ATP synthase and isocitrate lyase for maintenance of ATP homeostasis and viability. Microbiology 156, 81–87.

Ghodbane, R., Mba Medie, F., Lepidi, H., et al., 2014. Long-term survival of tuberculosis complex mycobacteria in soil. Microbiology 160, 496–501.

Glauert, A.M., Brieger, E.M., 1955. The electron-dense bodies of *Mycobacterium phlei*. J. Gen. Microbiol. 13, 310–317.

Goehring, N.W., Beckwith, J., 2005. Diverse paths to midcell: assembly of the bacterial cell division machinery. Curr. Biol. 15, R514–R526.

Goldman, D.S., 1970. Subcellular localization of individual mannose-containing phospholipids in *Mycobacterium tuberculosis*. Am. Rev. Respir. Dis. 102, 543–555.

Gomez, J.E., McKinney, J.D., 2004. *M. tuberculosis* persistence, latency, and drug tolerance. Tuberculosis (Edinb.) 84, 29–44.

Goodfellow, M., Wayne, L.G., 1982. Taxonomy and nomenclature. In: Ratledge, C., Stanford, J. (Eds.), The Biology of the Mycobacteria. Physiology, Identification and Classification, vol. 1. Academic Press, London & New York, pp. 471–521.

Goodfellow, M., Kampfer, P., Busse, H.J., et al., 2012. Taxonomic outline of the phylum actinobacteria. In: Whitman, W.B. (Ed.), Bergey's Manual of Systematic Bacteriology, second ed. Springer, New York, pp. 1–2024.

Grange, J.M., 1982. Koch's Tubercle Bacillus – a centenary reappraisal. Zentralbl. Bakteriol. Mikrobiol. Hyg. A 251, 297–307.

Grange, J.M., 1990. Drug-resistance and tuberculosis elimination. Bull. Int. Union Tuberc. Lung Dis. 65, 57–59.

Grange, J.M., 1998. Immunophysiology and immunopathology of tuberculosis. In: Davies, P.D.O. (Ed.), Clinical Tuberculosis. Chapman & Hall, London, pp. 129–152.

Grange, J.M., Yates, M.D., de Kantor, I.N., 1996. Guidelines for Speciation Within the *Mycobacterium tuberculosis* Complex. World Health Organization. Retrieved 2007.

Groenen, P.M., Bunschoten, A.E., Van Soolingen, D., et al., 1993. Nature of DNA polymorphism in the direct repeat cluster of *Mycobacterium tuberculosis*; application for strain differentiation by a novel typing method. Mol. Microbiol. 10, 1057–1065.

Grosset, J., 1980. Bacteriologic basis of short-course chemotherapy for tuberculosis. Clin. Chest Med. 1, 231–241.

Gutacker, M.M., Mathema, B., Soini, H., et al., 2006. Single-nucleotide polymorphism-based population genetic analysis of *Mycobacterium tuberculosis* strains from 4 geographic sites. J. Infect. Dis. 193, 121–128.

Hansen, G.A., 1880. Bacillus leprae. Virchows Arch. 79, 32–42.

Harada, K., 1976. The nature of mycobacterial acid-fastness. Stain Technol. 51, 255–260.

Harry, E.J., 2001. Bacterial cell division: regulating Z-ring formation. Mol. Microbiol. 40, 795–803.

Harry, E.J., Rodwell, J., Wake, R.G., 1999. Co-ordinating DNA replication with cell division in bacteria: a link between the early stages of a round of replication and mid-cell Z ring assembly. Mol. Microbiol. 33, 33–40.

Hartmans, S., De Bont, J.A.M., 1992. The genus Mycobacterium-nonmedical. In: Balows, A., Truper, H.G., Dworking, M., et al. (Eds.), The Prokaryotes. A Handbook on the Biology of Bacteria: Ecophysiology, Isolation, Identification, Applications, vol. 2. Springer Verlag, New York, pp. 1214–1237.

Hartmans, S., De Bont, J.A.M., Stackebrandt, E., 2006. The genus Mycobacterium—nonmedical. In: Balows, A., et al. (Ed.), The Prokaryotes, 3. second ed. Springer Verlag, New York, pp. 889–918.

Hernandez-Pando, R., Jeyanathan, M., Mengistu, G., et al., 2000. Persistence of DNA from *Mycobacterium tuberculosis* in superficially normal lung tissue during latent infection. Lancet 356, 2133–2138.

Hett, E.C., Rubin, E.J., 2008. Bacterial growth and cell division: a mycobacterial perspective. Microbiol. Mol. Biol. Rev. 72, 126–156.

Hirata, T., 1979. Electron microscopic observations of intracytoplasmic membrane systems and cell division in *Mycobacterium lepraemurium*. Int. J. Lepr. Other Mycobact. Dis. 47, 585–596.

Hobby, G.L., Auerbach, O., Lenert, T.F., et al., 1954. The late emergence of *M. tuberculosis* in liquid cultures of pulmonary lesions resected from humans. Am. Rev. Tuberc. 70, 191–218.

Hoffmann, C., Leis, A., Niederweis, M., et al., 2008. Disclosure of the mycobacterial outer membrane: cryo-electron tomography and vitreous sections reveal the lipid bilayer structure. Proc. Natl. Acad. Sci. U.S.A. 105, 3963–3967.

Honer, Z.U., Bentrup, K., Russell, D.G., 2001. Mycobacterial persistence: adaptation to a changing environment. Trends Microbiol. 9, 597–605.

Hosseini, H., Fooladi, A.A., Arjomandzadegan, M., et al., 2014. Genetics study and transmission electron microscopy of pili in susceptible and resistant clinical isolates of *Mycobacterium tuberculosis*. Asian Pac. J. Trop. Med. 7, 199–203.

Hruska, K., Kaevska, M., 2012. Mycobacteria in water, soil, plants and air: a review. Vet. Med. 57, 623–679.

Hu, Y.M., Butcher, P.D., Sole, K., et al., 1998. Protein synthesis is shutdown in dormant *Mycobacterium tuberculosis* and is reversed by oxygen or heat shock. FEMS Microbiol. Lett. 158, 139–145.

Imaeda, T., Convit, J., 1962. Electron microscope study of *Mycobacterium leprae* and its environment in a vesicular leprous lesion. J. Bacteriol. 83, 43–52.

Imaeda, T., Ogura, M., 1963. Formation of intracytoplasmic membrane system of mycobacteria related to cell division. J. Bacteriol. 85, 150–163.

Imaeda, T., Kirchheimer, W.F., Barksdale, L., 1982. DNA isolated from *Mycobacterium leprae*: genome size, base ratio, and homology with other related bacteria as determined by optical DNA-DNA reassociation. J. Bacteriol. 150, 414–417.

Jarlier, V., Nikaido, H., 1994. Mycobacterial cell wall: structure and role in natural resistance to antibiotics. FEMS Microbiol. Lett. 123, 11–18.

Joshi, J.M., 2011. Tuberculosis chemotherapy in the 21 century: back to the basics. Lung India 28, 193–200.

Joyce, G., Williams, K.J., Robb, M., et al., 2012. Cell division site placement and asymmetric growth in mycobacteria. PLoS One 7, e44582.

Juhlin, I., 1967. Contribution to the classification of mycobacteria and nocardias. Acta Pathol. Microbiol. Scand. 70 (Suppl. 189), 1.

Källenius, G., Koivula, T., Ghebremichael, S., et al., 1999. Evolution and clonal traits of *Mycobacterium tuberculosis* complex in Guinea-Bissau. J. Clin. Microbiol. 37, 3872–3878.

Kamerbeek, J., Schouls, L., Kolk, A., et al., 1997. Simultaneous detection and strain differentiation of *Mycobacterium tuberculosis* for diagnosis and epidemiology. J. Clin. Microbiol. 35, 907–914.

Kang, C.M., Nyayapathy, S., Lee, J.Y., et al., 2008. Wag31, a homologue of the cell division protein DivIVA, regulates growth, morphology and polar cell wall synthesis in mycobacteria. Microbiology 154, 725–735.

Kang, Y.A., Choi, N.K., Seong, J.M., et al., 2014. The effects of statin use on the development of tuberculosis among patients with diabetes mellitus. Int. J. Tuberc. Lung Dis. 18, 717–724.

Kaprelyants, A.S., Gottschal, J.C., Kell, D.B., 1993. Dormancy in nonsporulating bacteria. FEMS Microbiol. Rev. 10, 271–285.

Karakousis, P.C., Williams, E.P., Bishai, W.R., 2008. Altered expression of isoniazid-regulated genes in drug-treated dormant *Mycobacterium tuberculosis*. J. Antimicrob. Chemother. 61, 323–331.

Karlson, A.G., Lessel, E.F., 1970. Mycobacterium bovis nom. nov. Int. J. Syst. Bacteriol. 20, 273–282.

Kartmann, B., Stenger, S., Niederweis, M., 1999. Porins in the cell wall of *Mycobacterium tuberculosis*. J. Bacteriol. 181, 6543–6546.

Kazda, J.F., 1983. The principles of the ecology of mycobacteria. In: Ratledge, C., Stanford, J. (Eds.), The Biology of the Mycobacteria. Immunological and Environmental Aspects, vol. 2. Academic Press, London UK, pp. 323–415.

Keren, I., Minami, S., Rubin, E., et al., 2011. Characterization and transcriptome analysis of *Mycobacterium tuberculosis* persisters. mBio 2, e00100–11.

Khomenko, A.G., 1987. The variability of *Mycobacterium tuberculosis* in patients with cavitary pulmonary tuberculosis in the course of chemotherapy. Tubercle 68, 243–253.

Khomenko, A.G., Fadeeva, N.I., Golyshevskaya, V.I., 1983. Morphological and biochemical changes in *M. tuberculosis* during chemotherapy. Probl. Tuberc. 7, 48–53.

Kieser, K.J., Rubin, E.J., 2014. How sisters grow apart: mycobacterial growth and division. Nat. Rev. Microbiol. 12, 550–562.

Kim, S.J., Hong, Y.P., Lew, W.J., et al., 1995. Incidence of pulmonary tuberculosis among diabetics. Tuber. Lung Dis. 76, 529–533.

King, M.J., Park, W.H., 1929. Effect of Calmette's BCG vaccine on experimental animals. Am. J. Public Health Nations Health 19, 179–192.

Klemm, P., Schembri, M.A., 2000. Bacterial adhesins: function and structure. Int. J. Med. Microbiol. 290, 27–35.

Klopper, M., Warren, R.M., Hayes, C., et al., 2013. Emergence and spread of extensively and totally drug-resistant tuberculosis, South Africa. Emerg. Infect. Dis. 19, 449–455.

Kölbel, H., 1970. Das, substrat der variabilitat von Mykobakterien electronenmikroskipische bild. Pneumonology 142, 137–164.

Koch, R., 1882. The etiology of tuberculosis. Berlin Clin. Weekly 19, 221–230.

Koike, M., Takeya, K., 1961. Fine structures of intracytoplasmic organelles of mycobacteria. J. Biophys. Biochem. Cytol. 9, 597–608.

Konomi, N., Lebwohl, E., Mowbray, K., et al., 2002. Detection of mycobacterial DNA in Andean mummies. J. Clin. Microbiol. 40, 4738–4740.

Kotani, S., Kitaura, T., Hirano, T., et al., 1959. Isolation and chemical composition of the cells walls of BCG. Biken J. 2, 129–141.

Kremer, K., van Soolingen, D., Frothingham, R., et al., 1999. Comparison of methods based on different molecular epidemiological markers for typing of *Mycobacterium tuberculosis* complex strains: interlaboratory study of discriminatory power and reproducibility. J. Clin. Microbiol. 37, 2607–2618.

Kremer, L., de Chastellier, C., Dobson, G., et al., 2005. Identification and structural characterization of an unusual mycobacterial monomeromycolyl-diacylglycerol. Mol. Microbiol. 57, 1113–1126.

Kubica, G.P., 1973. Differential identification of mycobacteria: VII. Key features of identification of clinically significant mycobacteria. Am. Rev. Respir. Dis. 107, 9–21.

Kubica, G., Wayne, L.C., 1984. The Mycobacteria. Marcel Dekker Inc., New York.

Kumar, G., Kalra, V.K., Brodie, A.F., 1979. Asymmetric distribution of phospholipids in membranes from *Mycobacterium phlei*. Arch. Biochem. Biophys. 198, 22–30.

Kwapinski, J.B., Alcasid, A., Palser, H., 1970. Serological relationships of endoplasm antigens of saprophytic mycobacteria. Can. J. Microbiol. 16, 871–876.

Laqueyrerie, A., Militzer, P., Romain, F., et al., 1995. Cloning, sequencing, and expression of the apa gene coding for the *Mycobacterium tuberculosis* 45I47-kilodalton secreted antigen complex. Infect. Immun. 63, 4003–4010.

Lechevalier, M.P., Lechevalier, H.A., 1970. Chemical composition as a criterion in the classification of aerobic actinomycetes. Int. J. Syst. Bacteriol. 20, 435–443.

Lechevalier, M.P., Horan, A.C., Lechevalier, H., 1971. Lipid composition in the classification of nocardiae and mycobacteria. J. Bacteriol. 105, 313–318.

LeCoeur, H.F., Lagrange, P.H., Truffot-Pernot, C., et al., 1989. Relapses after stopping chemotherapy for experimental tuberculosis in genetically resistant and susceptible strains of mice. Clin. Exp. Immunol. 76, 458–462.

Lehmann, K.B., Neumann, R., 1896. In: Lehmann, J.F. (Ed.), Atlas und Grundriss der Bakteriologie und Lehrbuch der speciellen bakteriologischen Diagnostik (Munich).

Lévy-Frébault, V.V., Portaels, F., 1992. Roposed minimal standards for the genus Mycobacterium and for description of new slowly growing Mycobacterium species. Int. J. Syst. Bacteriol. 42, 315–323.

Li, L., Mendis, N., Trigui, H., et al., 2014. The importance of the viable but non-culturable state in human bacterial pathogens. Front. Microbiol. 5, 258.

Lillebaek, T., Dirksen, A., Baess, I., et al., 2002. Molecular evidence of endogenous reactivation of *Mycobacterium tuberculosis* after 33 years of latent infection. J. Infect. Dis. 185, 401–404.

Liu, T.Y., Gotschlich, E.C., 1967. Muramic acid phosphate as a component of the mucopeptide of Gram-positive bacteria. J. Biol. Chem. 242, 471–476.

Liu, J., Barry 3rd, C.E., Besra, G.S., et al., 1996. Mycolic acid structure determines the fluidity of the mycobacterial cell wall. J. Biol. Chem. 271, 29545–29551.

Loring, W.E., Vandiviere, H.M., 1956. The treated pulmonary lesion and its tubercle bacillus. I. Pathology and pathogenesis. Am. J. Med. Sci. 232, 20–29.

Lubarsch, O., 1899. Zur Kenntniss der Strahlenpilze. Z. Hyg. Infektionskrankh 31, 187–220.

Mahapatra, S., Yagi, T., Belisle, J.T., et al., 2005. Mycobacterial lipid II is composed of a complex mixture of modified muramyl and peptide moieties linked to decaprenyl phosphate. J. Bacteriol. 187, 2747–2757.

Malassez, L., Vignal, W., 1883. Sur le micro-organism de la tuberculose zooleique. Arch. Physiol. Norm Path III 4, 81–105.

Markova, N., Slavchev, G., Michailova, L., 2012. Unique biological properties of *Mycobacterium tuberculosis* L-form variants: impact for survival under stress. Int. Microbiol. 15, 61–68.

Martin, A., Portaels, F., 2007. Drug resistance and drug resistance detection. In: Palomino, J.C., Leão, S.C., Ritacco, V. (Eds.), Tuberculosis – from Basic Science to Patient Care, pp. 635–660 Available from: tuberculosistextbook.com.

Masjedi, M.R., Farnia, P., Sorooch, S., et al., 2006. Extensively drug-resistant tuberculosis: 2 years of surveillance in Iran. Clin. Infect. Dis. 43, 841–847.

Mattman, L.H., 1970. Cell wall deficient forms of Mycobacteria. Ann. N.Y. Acad. Sci. 174, 852–861.

Mattman, L.H., 2000. Cell Wall Deficient Forms, Third Edition: Stealth Pathogens. CRC Press, Boca Raton.

Mc Laughlin, A.M., O'Donnell, R.A., Gibbons, N., et al., 2007. Extensively drug-resistant tuberculosis (XDR-TB) – a potential threat in Ireland. Open Respir. Med. J. 1, 7–9.

McCarthy, C., 1971. Electronic counting in growth studies of *Mycobacterium avium*. Appl. Microbiol. 22, 546–551.

McCarthy, C., 1974. Effect of palmitic acid utilization on cell division in *Mycobacterium avium*. Infect. Immun. 9, 363–372.

McCarthy, C., 1976. Synthesis and release of sulfolipids by *Mycobacterium avium* during growth and cell division. Infect. Immun. 14, 1241–1252.

McCarthy, C., 1978. Ammonium ion requirement for the cell cycle of *Mycobacterium avium*. Infect. Immun. 19, 304–311.

McCready, K.A., Ratledge, C., 1978. Amounts of iron, haem and related compounds in *Mycobacterium smegmatis* grown in various concentrations of iron [proceedings]. Biochem. Soc. Trans. 6, 421–423.

McCune, R., Tompsett, R., 1956. Fate of *Mycobacterium tuberculosis* in mouse tissue as determined by the microbial enumeration technique. J. Exp. Med. 104, 737–762.

McCune, R.M., Tompsett, R., 1956. Fate of *Mycobacterium tuberculosis* in mouse tissues as determined by the microbial enumeration technique. II. The conversion of tuberculosis infection to the latent state by the administration of tuberculosis and a companion drug. J. Exp. Med. 104, 763–802.

McCune, R.M., Tompsett, R., McDermott, W., 1956. Fate of *Mycobacterium tuberculosis* in mouse tissues as determined by the microbial enumeration technique. I. The persistence of drug susceptible bacilli in the tissues despite prolonged antimicrobial therapy. J. Exp. Med. 104, 737–762.

McCune, R.M., Feldman, F.M., Lambert, H.P., et al., 1966. Microbial persistence II. Characteristics of the sterile state of tubercle bacilli. J. Exp. Med. 123, 469–486.

McDermott, W., 1958. Microbial persistence. Yale J. Biol. Med. 30, 257–291.

McDermott, W., 1959. Inapparent infection: relation of latent and dormant infection to microbial persistence. Pub. Health Rep. 74, 485–499.

McHugh, T.D., Batt, S.L., Shorten, R.J., et al., 2005. *Mycobacterium tuberculosis* lineage: a naming of the parts. Tuberculosis (Edinb.) 85, 127–136.

Medlar, E.M., Bernstein, S., Steward, D.M., 1952. A bacteriologic study of resected tuberculous lesions. Am. Rev. Tuberc. 66, 36–43.

Metschnikoff, E., 1888. Uber die phagocytare Rolle der Tuberkelriesenzellen. Virchow's Arch. Path Anat. 113, 63–94.

Michele, T.M., Ko, C., Bishai, W.R., 1999. Exposure to antibiotics induces expression of the *Mycobacterium tuberculosis* sigF gene: implications for chemotherapy against mycobacterial persistors. Antimicrob. Agents Chemother. 43, 218–225.

Migliori, G.B., De Iaco, G., Besozzi, G., et al., 2007. First tuberculosis cases in Italy resistant to all tested drugs. Euro Surveill. 12, E070517.1.

Minnikin, D.E., Polgar, N., 1966. Studies on the mycolic acids from human tubercle bacilli. Tetrahedron Lett. 23, 2643–2647.

Minnikin, D.E., 1982. Lipids: complex lipids, their chemistry, biosynthesis and roles. In: Ratledge, C., Stanford, J. (Eds.), The Biology of the Mycobacteria. Physiology, Identification and Classification, vol. 1. Academic Press, London, pp. 95–184.

Minnikin, D.E., Minnikin, S.M., Parlett, J.H., et al., 1984. Mycolic acid patterns of some species of Mycobacterium. Arch. Microbiol. 139, 225–231.

Misaki, A., Seto, N., Azuma, I., 1974. Structure and immunological properties of D-arabino-D-gdactan isolated from cell walls of Mycobacterium species. J. Biochem. 76, 15–27.

Mitchison, D.A., 1979. Basic mechanisms of chemotherapy. Chest 76, 771S–775S.

Mitchison, D.A., 1985. The action of antituberculosis drugs in shortcourse chemotherapy. Tubercle 66, 219–225.

Mitchison, D.A., 1998. How drug resistance emerges as a result of poor compliance during short course chemotherapy for tuberculosis. Int. J. Tuberc. Lung Dis. 2, 10–15.

Mitchison, D.A., Dickinson, J.M., 1978. Bactericidal mechanisms in short-course chemotherapy. Bull. Int. Union Tuberc. 53, 254–259.

Mitchison, D.A., Selkon, J.B., 1957. Bacteriological aspects of a survey of the incidence of drug-resistant tubercle bacilli among untreated patients. Tubercle 38, 85–98.

Mochizuki, M., 1991. Kinetics of oxygen and carbon dioxide reactions. In: Crystal, R.G., West, J.B., et al. (Eds.), The Lung: Scientific Foundations. Raven Press Ltd., New York, pp. 1241–1252.

Montoro, E., Rodriguez, R., 2007. Global burden of tuberculosis. In: Palomino, J.C., Leao, S.C., Ritacco, V. (Eds.), Tuberculosis 2007: From Basic Science to Patient Care. Amedeo Flying Publisher, pp. 263–282.

Morita, R.Y., 1993. Bioavailability of energy and the starvation state. In: Kjelleberg, S. (Ed.), Starvation in Bacteria. Plenum Press, New York, pp. 1–24.

Mostowy, S., Behr, M.A., 2002. Comparative genomics in the fight against tuberculosis: diagnostics, epidemiology, and BCG vaccination. Am. J. Pharmacogenomics 2, 189–196.

Mudd, S., Wintersheid, L.C., 1953. A note concerning bacterial cell walls, mitochondria and nuclei. Exp. Cell Res. 5, 251–254.

Mukamolova, G.V., Kaprelyants, A.S., Young, D.I., et al., 1998. A bacterial cytokine. Proc. Natl. Acad. Sci. U.S.A. 95, 8916–8921.

Murohasi, T., 1948. Reaction to tuberculin hypodermic tests of children who have had oral inoculation with B C G. Acta Pathol. Microbiol. Scand. 25, 15.

Nachega, J.B., Chaisson, R.E., 2003. Tuberculosis drug resistance: a global threat. Clin. Infect. Dis. 36, S24–S30.

Nakamura, S., 1958. The staining characteristics of resistant tubercle bacilli and the relation of morphological changes to the clinical picture. II. Electron microscopic morphology of resistant tubercle bacilli. Kekkaku 33, 499–505.

Nassal, J., Breunig, W., Schnedelbach, U., 1974. Atypical mycobacteria in fruit, vegetables, and cereals (in German). Praxis der Pneumologie 28, 667–674.

Nassau, E., Hamilton, G.M., 1957. Atypical Mycobacteria in human pulmonary disease. Tubercle 38, 387–396.

Niederweis, M., 2003. Mycobacterial porins–new channel proteins in unique outer membranes. Mol. Microbiol. 49, 1167–1177.

Niederweis, M., 2008. Nutrient acquisition by mycobacteria. Microbiology 154, 679–692.

Nikaido, H., 1994a. Prevention of drug access to bacterial targets: permeability barriers and active efflux. Science 264, 382–388.

Nikaido, H., 1994b. Diffusion of inhibitors across the cell wall. In: Ghuysen, J.M., Hakenbeck, R. (Eds.), Bacterial Cell Wall. Elsevier, Netherlands, Amsterdam, pp. 547–558.

Nikaido, H., 2003. Molecular basis of bacterial outer membrane permeability revisited. Microbiol. Mol. Biol. Rev. 67, 593–656.

Nikaido, H., Rosenberg, E.Y., 1981. Effect on solute size on diffusion rates through the transmembrane pores of the outer membrane of *Escherichia coli*. J. Gen. Physiol. 77, 121–135.

Nocard, M.E., Roux, E., 1887. Sur la culture du bacilli de la tuberculose. Ann. Inst. Pasteur (Paris) 1, 19–29.

Nyka, W., 1974. Studies on the effect of starvation on Mycobacteria. Infect. Immun. 9, 843–850.

O'Brien, D.J., Schmitt, S.M., Fitzgerald, S.D., Berry, D.E., 2011. Management of bovine tuberculosis in Michigan wildlife: current status and near term prospects. Vet. Microbiol. 151, 179–187.

Oka, S., Fukushi, K., Fujimoto, M., et al., 1968. La distribution subcellulaire des phospholipides de la mycobacthie. C.R. Soc. Frunco-Japon. Biol. 162, 1648–1650.

Opie, E.L., Aronson, J.D., 1927. Tubercle bacilli in latent tuberculous lesions and in lung tissue without tuberculous lesions. Arch. Pathol. 4, 1–21.

Ortalo-Magné, A., Dupont, M.A., Lemassu, A., et al., 1995. Molecular composition of the outermost capsular material of the tubercle bacillus. Microbiology 141, 1609–1620.

Ortalo-Magné, A., Andersen, A.B., Daffé, M., 1996. The outermost capsular arabinomannans and other mannoconjugates of virulent and avirulent tubercle bacilli. Microbiology 142, 927–935.

Ottmani, S.E., Murray, M.B., Jeon, C.Y., et al., 2010. Consultation meeting on tuberculosis and diabetes mellitus: meeting summary and recommendations. Int. J. Tuberc. Lung Dis. 12, 1513–1517.

Papapetropoulou, M., Tsintzou, A., Vantarakis, A., 1997. Environmental mycobacteria in bottled table waters in Greece. Can. J. Microbiol. 43, 499–502.

Parrish, N.M., Dick, J.D., Bishai, W.R., 1998. Mechanisms of latency in mycobacterium tuberculosis. Trends Microbiol. 107, 110–112.

Paul, T.R., Beveridge, T.J., 1994. Preservation of surface lipids and determination of ultrastructure of *Mycobacterium kansasii* by freeze-substitution. Infect. Immun. 62, 1542–1550.

Pealing, L., Wing, K., Mathur, R., et al., 2015. Risk of tuberculosis in patients with diabetes: population based cohort study using the UK Clinical Practice Research Datalink. BMC Med. 13, 135.

Petit, J.F., Adam, A., Wietzerbin-Falszpan, J., et al., 1969. Chemical structure of the cell wall of Mycobacterium smegmatis. I – isolation and partial characterization of the peptidoglycan. Biochem. Biophys. Res. Commun. 35, 478–485.

Phillips, C.J.C., Foster, C.R.W., Morris, P.A., et al., 2003. The transmission of *Mycobacterium bovis* infection to cattle. Res. Vet. Sci. 74, 1–15.

Picard, B., Frehel, C., Rastogi, N., 1984. Cytochemical characterization of mycobacterialouter surfaces. Acta Leprol. 95, 227–235.

Pitulle, C., Dorsch, M., Kazda, J., et al., 1992. Phylogeny of rapidly growing members of the genus Mycobacterium. Int. J. Syst. Bacteriol. 42, 337–343.

Porter, K.R., Yegian, D., 1945. Some artifacts encountered in stained preparations of tubercle bacilli. J. Bacteriol. 50, 563–575.

Price, D.S., 1949. The need for BCG vaccination in infants. Tubercle 30, 11–13.

Pyle, M.M., 1947. Relative numbers of resistant tubercle bacilli in sputa of patients before and during treatment with streptomycin. Proc. Staff Meet Mayo Clin. 22, 465–473.

Ramakrishnan, T., Murthy, P.S., Gopinathan, K.P., 1972. Intermediary metabolism of mycobacteria. Bacteriol. Rev. 36, 65–108.

Rastogi, N., 1991. 7th forum in microbiology – structure and functions of the cell envelope in relation to mycobacterial virulence, pathogenicity and multiple drug-resistance. Res. Microbiol. 142, 419–481.

Rastogi, N., 1993. Mycobacteria as intracellular pathogens: current notions of pathogenicity, virulence, and drug resistance and their relationship to effective therapy. In: Raoult, D. (Ed.), Antimicrobial Agents and Intracellular Pathogens. CRC Press, Boca Raton. Florida, pp. 245–300.

Rastogi, N., Barrow, W.W., 1994. 11th Forum in microbiology: laboratory and clinical aspects of the *Mycobacterium avium* epidemic: contributing factors associated with variability of drug susceptibility and immune responsiveness, and the multifaceted nature of pathogenicity. Res. Microbiol. 145, 167–261.

Rastogi, N., David, H.L., 1981. Growth and cell division of *Mycobacterium avium*. J. Gen. Microbiol. 126, 77–84.

Rastogi, N., David, H.L., 1993. Mode of action of antituberculous drugs and mechanisms of resistance in *Mycobacterium tuberculosis*. Res. Microbiol. 144, 133–143.

Rastogi, N., Sola, C., 2007. Molecular evolution of the *Mycobacterium tuberculosis* complex. In: Palomino, J.C., Leao, S.C., Ritacco, V. (Eds.), Tuberculosis 2007: From Basic Science to Patient Care. Amedeo Flying Publisher, pp. 53–91.

Rastogi, N., Fréhel, C., David, H.L., 1986. Triple-layered structure of mycobacterial cell wall: evidence for the existence of a polysaccharide-rich outer layer in eighteen mycobacterial species. Curr. Microbiol. 13, 237–242.

Rastogi, N., Legrand, E., Sola, C., 2001. The mycobacteria: an introduction to nomenclature and pathogenesis. Rev. Sci. Tech. 20, 21–54.

Raynaud, C., Papavinasasundaram, K.G., Speight, R.A., et al., 2002. The functions of OmpATb, a pore-forming protein of *Mycobacterium tuberculosis*. Mol. Microbiol. 46, 191–201.

Ridell, M., Norlin, M., 1973. Serological study of nocardia by using mycobacterial precipitation reference systems. J. Bacteriol. 113, 1–7.

Rieder, H.L., 2002. Interventions for Tuberculosis Control and Elimination. International Union Against Tuberculosis and Lung Diseases, Paris.

Rogall, T., Flohr, T., Böttger, E.C., 1990. Differentiation of mycobacterium species by direct sequencing of amplified DNA. J. Gen. Microbiol. 136, 1915–1920.

Romain, F., Laqueyrerie, A., Militzer, P., 1993. Identification of a *Mycobacterium bovis* BCG 45/47-kilodalton antigen complex, an immunodominant target for antibody response after immunization with living bacteria. Infect. Immun. 61, 742–750.

Rosenblatt, M.B., Fullam, E.F., Gessler, A.E., 1942. Studies of mycobacteria with the electron microscope. Am. Rev. Tuberc. 46, 587–599.

Roy, A., Eisenhut, M., Harris, R.J., et al., 2014. Effect of BCG vaccination against *Mycobacterium tuberculosis* infection in children: systematic review and meta-analysis. BMJ 349, g4643.

Runyon, E.H., 1959. Anonymous mycobacteria in pulmonary disease. Med. Clin. North Am. 43, 273–290.

Runyon, E.H., Selin, M.J., Harris, H.W., 1959. Distinguishing mycobacteria by the niacin test; a modified procedure. Am. Rev. Tuberc. 79, 663–665.

Runyon, E.H., Wayne, L.G., Kubica, G.P., 1974. Family II. Mycobacteriaceae Chester 1897, 63. p. 681–701. In: Buchanan, R.E., Gibbons, N.E. (Eds.), Bergey's Manual of Determinative Biology, eighth ed. Williams and Wilkins Co., Baltimore.

Ruslami, R., Aarnoutse, R.E., Alisjahbana, B., et al., 2010. Implications of the global increase of diabetes for tuberculosis control and patient care. Trop. Med. Int. Health 12, 1289–1299.

Russell, A.D., 1969. The mechanism of action of some antibacterial agents. In: Ellis, G.P., West, G.B. (Eds.), Progress in Medicinal Chemistry. Elsevier, New York, USA.

Samper, S., Martín, C., 2007. Spread of extensively drug-resistant tuberculosis. Emerg. Infect. Dis. 13, 647–648.

Santha, T., Nazareth, O., Krishnamurthy, S., et al., 1989. Treatment of pulmonary tuberculosis with short course chemotherapy in South India—5-year follow-up. Tubercle 70, 229–234.

Sarathy, J.P., Dartois, V., Lee, E.J.D., 2012. The role of transport mechanisms in Mycobacterium tuberculosis drug resistance and tolerance. Pharmaceuticals 5, 1210–1235.

Sarathy, J., Dartois, V., Dick, T., Gengenbacher, M., 2013. Reduced drug uptake in phenotypically resistant nutrient-starved nonreplicating *Mycobacterium tuberculosis*. Antimicrob. Agents Chemother. 57, 1648–1653.

Sarathy, J.P., Dartois, V., Dick, T., 2013a. Impaired drug uptake contributes to phenotypic resistance in nutrient-starved non-replicating *Mycobacterium tuberculosis*. Antimicrob. Agents Chemother. 57, 1648–1653.

Sarathy, J.P., Lee, E.J.D., Dartois, V., 2013b. Polyamines inhibit fluoroquinolone uptake in mycobacteria. PLoS One 8, e65806.

Saviola, B., Bishai, W., 2006. The genus Mycobacterium—Medical. Prokaryotes 3, 919–933.

Schaefer, W.B., Lewis Jr., C.W., 1965. Effect of oleic acid on growth and cell structure of mycobacteria. J. Bacteriol. 90, 1438–1447.

Sebban, M., Mokrousov, I., Rastogi, N., et al., 2002. A data-mining approach to spacer oligonucleotide typing of *Mycobacterium tuberculosis*. Bioinformatics 18, 235–243.

Senaratne, R.H., Mobasheri, H., Papavinasasundaram, K.G., et al., 1998. Expression of a gene for a porin-like protein of the OmpA family from *Mycobacterium tuberculosis* H37Rv. J. Bacteriol. 180, 3541–3547.

Shah, N.S., Wright, A., Bai, G.H., et al., 2007. Worldwide emergence of extensively drug-resistant tuberculosis. Emerg. Infect. Dis. 13, 380–387.

Shean, K.P., Willcox, P.A., Siwendu, S.N., et al., 2008. Treatment outcome and follow-up of multidrug-resistant tuberculosis patients, West Coast/Winelands, South Africa, 1992–2002. Int. J. Tuberc. Lung Dis. 12, 1182–1189.

Shimao, T., 1987. Drug-resistance in tuberculosis control. Tubercle 68, 5–18.

Shinnick, T.M., Good, R.C., 1994. Mycobacterial taxonomy. Eur. J. Clin. Microbiol. Infect. Dis. 13, 884–901.

Shleeva, M.O., Bagramyan, K., Telkov, M.V., et al., 2002. Formation and resuscitation of 'non-culturable' cells of Rhodococcus rhodochrous and *Mycobacterium tuberculosis* in prolonged stationary phase. Microbiology 148, 1581–1591.

Shleeva, M.O., Kudykina, Y.K., Vostroknutova, G.N., et al., 2011. Dormant ovoid cells of *Mycobacterium tuberculosis* are formed in response to gradual external acidification. Tuberculosis (Edinb.) 91, 146–154.

Sieniawska, E., Swatko-Ossor, M., Sawicki, R., et al., 2015. Morphological changes in the overall *Mycobacterium tuberculosis* H37Ra cell shape and cytoplasm homogeneity due to mutellina purpurea *L.* essential oil and its main constituents. Med. Princ. Pract. 24, 527–532.

Silva, M.T., Macedo, P.M., 1982. Ultrastructure of *Mycobacterium leprae* and other acid-fast bacteria as influenced by fixation conditions. Ann. Microbiol. (Paris) 133, 59–73.

Singh, B., Ghosh, J., Islam, N.M., et al., 2010. Growth, cell division and sporulation in mycobacteria. Antonie Van Leeuwenhoek 98, 165–177.

Singh, B., Nitharwal, R.G., Ramesh, M., et al., 2013. Asymmetric growth and division in Mycobacterium spp.: compensatory mechanisms for non-medial septa. Mol. Microbiol. 88, 64–76.

Smeulders, M.J., Keer, J., Speight, R.A., et al., 1999. Adaptation of *Mycobacterium smegmatis* to stationary phase. J. Bacteriol. 181, 270–283.

Smida, J., Kazda, J., Stackebrandt, E., 1988. Molecular evidence for the relationship of *Mycobacterium leprae* to slow-growing pathogenetic mycobacteria. Int. J. Lepr. 56, 449–454.

Smith, I., 2004. What is the health, social, and economic burden of tuberculosis. In: Frieden, T. (Ed.), Toman's Tuberculosis Case Detection, Treatment, and Monitoring: Questions and Answers, second ed. WHO, Geneva, pp. 233–237.

Sneath, P.H.A., 1970. In: Prauser, H. (Ed.), The Actinomycetales. Fischer, Jena, pp. 371–377.

Sola, C., Horgen, L., Goh, K.S., et al., 1997. Mulecular fingerprinting of *Mycobacterium tuberculosis* on a Caribbean island with IS6110 and Drr probes. J. Clin. Microbiol. 35, 843–846.

Sola, C., Devallois, A., Horgen, L., et al., 1999. Tuberculosis in the Caribbean: using spacer oligonucleotide typing to understand strain origin and transmission. Emerg. Infect. Dis. 5, 404–414.

Sola, C., Filliol, I., Gutierrez, M.C., et al., 2001. Spoligotype database of mycobacterium tuberculosis: biogeographic distribution of shared types and epidemiologic and phylogenetic perspectives. Emerg. Infect. Dis. 7, 390–396.

Sreevatsan, S., Pan, X., Stockbauer, K.E., et al., 1997. Restricted structural gene polymorphism in the *Mycobacterium tuberculosis* complex indicates evolutionarily recent global dissemination. Proc. Natl. Acad. Sci. U.S.A. 94, 9869–9874.

Stackebrandt, E., Ludwig, W., Weizenegger, M., et al., 1987. Comparative 16S rRNA oligonucleotide analyses and murein types of round-spore-forming bacilli and non-spore-forming relatives. J. Gen. Microbiol. 133, 2523–2529.

Stackebrandt, E., Rainey, F.A., Ward-Rainey, N.L., 1997. Proposal for a new hierarchic classification system, Actinobacteria, classis nov. Int. J. Syst. Bacteriol. 47, 479–491.

Stahl, D.A., Urbance, J.W., 1990. The division between fast- and slow-growing species corresponds to natural relationships among the mycobacteria. J. Bacteriol. 172, 116–124.

Stewart, E.J., Madden, R., Paul, G., et al., 2005. Aging and death in an organism that reproduces by morphologically symmetric division. PLoS Biol. 3, e45.

Streicher, E.M., Warren, R.M., Kewley, C., et al., 2004. Genotypic and phenotypic characterization of drug-resistant *Mycobacterium tuberculosis* isolates from rural districts of the Western Cape Province of South Africa. J. Clin. Microbiol. 42, 891–894.

Strom, M.S., Lory, S., 1993. Structure-function and biogenesis of the type IV pili. Annu. Rev. Microbiol. 47, 565–596.

Supply, P., Lesjean, S., Savine, E., et al., 2001. Automated high throughput genotyping for study of global epidemiology of *Mycobacterium tuberculosis* based on mycobacterial interspersed repetitive units. J. Clin. Microbiol. 39, 3563–3571.

Sweany, H.C., Levinson, S.A., Stadnichenko, A.M.S., 1943. Tuberculous infection in people dying of causes other than tuberculosis. Am. Rev. Tuberc. 47, 131–173.

Takade, A., Takeya, K., Taniguchi, H., et al., 1983. Electron microscopic observations of cell division in *Mycobacterium vaccae* V1. J. Gen. Microbiol. 129, 2315–2320.

Takahashi, S.L., 1979. Phase growth of mycobacteria. II. Consideration on the survival of tubercle bacillus in the caseous lesion. Kekkaku 54, 231–236.

Tárnok, I., Tárnok, Z., 1970. Carotene and Xanthophylls in mycobacteria.I. Technical Procedures thin- layer chromatographic patterns of mycobacterial pigments. Tubercle 51, 305–312.

Taylor, G.M., Young, D.B., Mays, S.A., 2005. Genotypic analysis of the earliest known prehistoric case of tuberculosis in Britain. J. Clin. Microbiol. 43, 2236–2240.

Thanky, N.R., Young, D.B., Robertson, B.D., 2007. Unusual features of the cell cycle in Mycobacteria: polar-restricted growth and the snapping-model of cell division. Tubercle 87, 231–236.

Tomasz, A., Albino, A., Zanati, E., 1970. Multiple antibiotic resistance in a bacterium with suppressed autolytic system. Nature 227, 138–140.

Torrea, G., Levee, G., Grimont, P., et al., 1955. Chromosomal DNA fingerprinting analysis using the insertion sequence IS6110 and the repetitive element DR as strain-specific markers for epidemiological study of tuberculosis in French Polynesia. J. Clin. Microbiol. 33, 1899–1904.

Trevors, J.T., 2004. Evolution of cell division in bacteria. Theory Biosci. 123, 3–15.

Trias, J., Benz, R., 1993. Characterization of the channel formed by the mycobacterial porin in lipid bilayer membranes. Demonstration of voltage gating and of negative point charges at the channel mouth. J. Biol. Chem. 268, 6234–6240.

Trias, J., Jarlier, V., Benz, R., 1992. Porins in the cell wall of mycobacteria. Science 258, 1479–1481.

Trnka, L., Wiegeshaus, E., Smith, D.W., 1968. Resolution and identification of ribosomes from mycobacteria. J. Bacteriol. 95, 310–313.

Trunz, B.B., Fine, P., Dye, C., 2006. Effect of BCG vaccination on childhood tuberculous meningitis and miliary tuberculosis worldwide: a meta-analysis and assessment of cost-effectiveness. Lancet 367, 1173–1180.

Tsolaki, A.G., Gagneux, S., Pym, A.S., et al., 2005. Genomic deletions classify the Beijing/W strains as a distinct genetic lineage of *Mycobacterium tuberculosis*. J. Clin. Microbiol. 43, 3185–3191.

Tuomanen, E., 1986. Phenotypic tolerance: the search for betalactam antibiotics that kill nongrowing bacteria. Rev. Infect. Dis. 8, S279–S291.

Typas, A., Banzhaf, M., van den Berg van Saparoea, B., et al., 2010. Regulation of peptidoglycan synthesis by outer-membrane proteins. Cell 143, 1097–1109.

Udwadia, Z.F., Amale, R.A., Ajbani, K.K., et al., 2012. Totally drug-resistant tuberculosis in India. Clin. Infect. Dis. 54, 579–581.

Uhía, I., Williams, K.J., Shahrezaei, V., et al., 2015. Mycobacterial growth. Cold Spring Harb. Perspect. Med. 5, 10.

Ulrich, T., Kaufmann, S.H.E., 2004. Cell-mediated immune response. In: Rom, W.N., Garay, S.M. (Eds.), Tuberculosis. Lippincott Williams & Wilkins, Philadelphia, pp. 251–262.

Van Ingen, J., Rahim, Z., Mulder, A., et al., 2012. Characterization of *Mycobacterium orygis* as *M. tuberculosis* complex subspecies. Emerg. Infect. Dis. 18, 653–655.

Van Soolingen, D., Qian, L., de Haas, P.E., et al., 1995. Predominance of a single genotype of *Mycobacterium tuberculosis* in countries of east Asia. J. Clin. Microbiol. 33, 3234–3238.

Van Soolingen, D., Hoogenboezem, T., de Haas, P.E., et al., 1997. A novel pathogenic taxon of the *Mycobacterium tuberculosis* complex, Canetti: characterization of an exceptional isolate from Africa. Int. J. Syst. Bacteriol. 47, 1236–1245.

Van Soolingen, D., Kremer, K., Hermans, P.W.M., 2007. Molecular epidemiology: Breakthrough achievements and future prospects. In: Palomino, J.C., Leao, S.C., Ritacco, V. (Eds.), Tuberculosis 2007: From Basic Science to Patient Care. Amedeo Flying Publisher, pp. 315–339.

Vareldzis, B.P., Grosset, J., de Kantor, I., et al., 1994. Drug-resistant tuberculosis: laboratory issues. World Health Organization recommendations. Tuber. Lung Dis. 75, 1–7.

Velayati, A.A., Farnia, P., 2012a. Morphological characterization of *Mycobacterium tuberculosis*. In: Cardona, P.J. (Ed.), Understanding Tuberculosis – Deciphering the Secret Life of the Bacilli.

Velayati, A.A., Farnia, P., 2012b. Division-cycle in Mycobacterium. Int. J. Mycobacteriol. 1, 111–117.

Velayati, A.A., Farnia, P., Boloorsaze, M.R., et al., 2007. *Mycobacterium bovis* infection in children in the same family: transmission through inhalation. Monaldi Arch. Chest Dis. 67, 169–172.

Velayati, A.A., Farnia, P., Ibrahim, T.A., et al., 2009a. Differences in cell wall thickness between resistant and nonresistant strains of *Mycobacterium tuberculosis*: using transmission electron microscopy. Chemotherapy 55, 303–307.

Velayati, A.A., Farnia, P., Masjedi, M.R., et al., 2009b. Totally drug-resistant tuberculosis strains: evidence of adaptation at the cellular level. Eur. Respir. J. 34, 1202–1203.

Velayati, A.A., Masjedi, M.R., Farnia, P., et al., 2009c. Emergence of new forms of totally drug-resistant tuberculosis bacilli: super extensively drug-resistant tuberculosis or totally drug-resistant strains in Iran. Chest 136, 420–425.

Velayati, A.A., Farnia, P., Merza, M.A., et al., 2010. New insight into extremely drug-resistant tuberculosis: using atomic force microscopy. Eur. Respir. J. 36, 1490–1493.

Velayati, A.A., Farnia, P., Masjedi, M.R., et al., 2011. Sequential adaptation in latent tuberculosis bacilli: observation by atomic force microscopy (AFM). Int. J. Clin. Exp. Med. 4, 193–199.

Velayati, A.A., Farnia, P., Masjedi, M.R., et al., 2012a. Morphological modification by Tubercle bacilli: no time for denial. J. Infect. Dev. Ctries 6, 97–99.

Velayati, A.A., Farnia, P., Masjedi, M.R., 2012b. Pili in totally drug resistant Mycobacterium Tuberculosis (TDR-TB). IJMYCO 1, 57–58.

Velayati, A.A., Farnia, P.P., Masjedi, M.R., 2012c. Recurrence after treatment success in pulmonary multidrug-resistant tuberculosis: predication by continual PCR positivity. Int. J. Clin. Exp. Med. 5, 271–272.

Velayati, A.A., Farnia, P., Masjedi, M.R., 2013. The totally drug resistant tuberculosis (TDR-TB). Int. J. Clin. Exp. Med. 6, 307–309.

Velayati, A.A., Abeel, T., Shea1, T., et al., 2016. Populations of latent *Mycobacterium tuberculosis* lack a cell wall: isolation, visualization and whole genomic characterization. IJYMCO. 5, 59–65.

Vera, H.D., Rettger, L.F., 1940. Morphological variations of the tubercle bacillus and certain recently isolated soil acid fasts with emphasis on filterability. J. Bacteriol. 39, 659–687.

Vestal, A.L., Kubica, G.P., 1966. Differential colonical characteristics of mycobacteria on Middlebrook and cohn 7H11 agar-base medium. Am. Rev. Respir. Dis. 94, 247–252.

Veyron-Churlet, R., Brust, B., Kremer, L., et al., 2011. Expression of OmpATb is dependent on small membrane proteins in *Mycobacterium bovis* BCG. Tuberculosis (Edinb.) 91, 544–548.

Vignal, W., 1882. Some remarks on Ehrlich and Gibbes's methods for the detection of *Bacillus tuberculosis*. Br. Med. J. 826.

Vijay, S., Nagaraja, M., Sebastian, J., et al., 2014. Asymmetric cell division in *Mycobacterium tuberculosis* and its unique features. Arch. Microbiol. 196, 157–168.

Vincent, V., 1994. Les mycobactéries: bacilles de la tuberculose. In: Freney, J., Renaud, F., Hansen, W., et al. (Eds.), Manuel de bactériologie clinique, vol. 2. Elsevier, Paris, pp. 899–922.

Wagner, J.C., Bokkenheuser, V., 1961. The mycobacterium isolated from the dassie Procavia capensis (Pallas). Tubercle 42, 47–56.

Waksman, S.A., 1964. The Conquest of Tuberculosis. Robert Hale Limited, London.

Wakamoto, Y., Dhar, N., Chait, R., et al., 2013. Dynamic persistence of antibiotic-stressed mycobacteria. Science 4, 91–95.

Wallis, R.S., Patil, S., Cheon, S.H., et al., 1999. Drug tolerance in *Mycobacterium tuberculosis*. Antimicrob. Agents Chemother. 43, 2600–2606.

Wang, J.Y., Lee, L.N., Hsueh, P.R., 2005. Factors changing the manifestation of pulmonary tuberculosis. Int. J. Tuberc. Lung Dis. 9, 777–783.

Wayne, L.G., 1956. The bacteriology of resected tuberculous pulmonary lesions. I. The effect of interval between reversal of infectiousness and subsequent surgery. Am. Rev. Tuberc. Pulm. Dis. 74, 376–387.

Wayne, L.G., 1976. Dynamics of submerged growth of *Mycobacterium tuberculosis* under aerobic and microaerophilic conditions. Am. Rev. Respir. Dis. 114, 807–811.

Wayne, L.G., 1978. Mycobacterial taxonomy: a search for discontinuities. Ann Microbiol. 129, 13–27.

Wayne, L.G., 1981. Numerical taxonomy and cooperative studies: roles and limits. Rev. Infect. Dis. 3, 822–828.

Wayne, L.G., 1994. Dormancy of *Mycobacterium tuberculosis* and latency of disease. Eur. J. Clin. Microbiol. Infect. Dis. 13, 908–914.

Wayne, L.G., Diaz, G.A., 1985. Identification of mycobacteria by specific precipitation of catalase with absorbed sera. J. Clin. Microbiol. 21, 721–725.

Wayne, L.G., Hayes, L.G., 1996. An in vitro model for sequential study of shiftdown of *Mycobacterium tuberculosis* through two stages of nonreplicating persistence. Infect. Immun. 64, 2062–2069.

Wayne, L.G., Kubica, G.P., 1986. Genus Mycobacterium Lehmann and Neumann 1896, 363. In: Sneath, P.H.A., Mair, N.S., Sharpe, M.E., et al. (Eds.), Bergey's Manual of Systematic Bacteriology. Williams & Wilkins, United States, pp. 1436–1457.

Wayne, L.G., Lin, K.Y., 1982. Glyoxylate metabolism and adaptation of *Mycobacterium tuberculosis* to survival under anaerobic conditions. Infect. Immun. 37, 1042–1049.

Wayne, L.G., Sohaskey, C.D., 2001. Nonreplicating persistence of mycobacterium tuberculosis. Annu. Rev. Microbiol. 55, 139–163.

Wayne, L.G., Sramek, H.A., 1994. Metronidazole is bactericidal to dormant cells of *Mycobacterium tuberculosis*. Antimicrob. Agents Chemother. 38, 2054–2058.

Wayne, L.G., Dietz, T.M., Gernez-Rieux, C., Jenkins, P.A., et al., 1971. A cooperative numerical analysis of scotochromogenic slowly growing mycobacteria. J. Gen. Microbiol. 66, 255–271.

Wayne, L.G., Good, R.C., Böttger, E.C., et al., 1996. Semantide- and chemotaxonomy-based analyses of some problematic phenotypic clusters of slowly growing mycobacteria, a cooperative study of the International Working Group on Mycobacterial Taxonomy. Int. J. Syst. Bacteriol. 46, 280–297.

Wejnert, C., Hess, K.L., Rose, C.E., et al., 2015. Age-specific race and ethnicity disparities in HIV infection and awareness among men who have sex with men-20 U.S. cities, 2008 to 2014. J. Infect. Dis. pii:jiv500.

Werner, G.H., 1951. Electron-microscopic studies on the cellular morphology of tubercle bacilli. Bibl. Tuberc. 5, 53–90.

Weyer, K., Mirzayev, F., Migliori, G.B., et al., 2013. Rapid molecular TB diagnosis: evidence, policy making and global implementation of Xpert MTB/RIF. Eur. Respir. J. 42, 252–271.

Wheate, H., 1951. A note on the morphological changes in the *Mycobacterium leprae* under sulphone therapy. Lepr. Rev. 22, 79–80.

Whiley, H., Keegan, A., Giglio, S., 2012. *Mycobacterium avium* complex – the role of potable water in disease transmission. J. Appl. Microbiol. 113, 223–232.

Whitman, W., Goodfellow, M., Kämpfer, P., et al., 2012. In: Bergey's Manual of Systematic Bacteriology, second ed. Springer, New York.

Wietzerbin, J., Das, B.C., Petit, J.F., et al., 1974. Occurrence of D-alanyl-(D)-meso-diaminopimeliaccid and meso diaminopimelyl-mesodiaminopimelic acid interpeptide linkages in the peptidoglycan of Mycobacteria (sic). Biochemistry 13, 3471–3476.

Wietzerbin-Falszpan, J., Das, B.C., Azuma, I., et al., 1970. Isolation and mass spectrometric identification of the peptide subunits of mycobacterial cell walls. Biochem. Biophys. Res. Commun. 40, 57–63.

Wilson, T.M., Collins, D.M., 1996. ahpC, a gene involved in isoniazid resistance of the *Mycobacterium tuberculosis* complex. Mol. Microbiol. 19, 1025–1034.

Wilson, M., DeRisi, J., Kristensen, H.H., et al., 1999. Exploring drug induced alterations in gene expression in *Mycobacterium tuberculosis* by microarray hybridization. Proc. Natl. Acad. Sci. U.S.A. 96, 12833–12838.

Wirth, T., Hildebrand, F., Allix-Béguec, C., et al., 2008. Origin, spread and demography of the *Mycobacterium tuberculosis* complex. PLoS Pathog. 4, e1000160.

Worcel, A., Goldman, D.S., Sachs, I.B., 1968. Properties and fine structure of the ribosomes from *Mycobacterium tuberculosis*. Proc. Natl. Acad. Sci. U.S.A. 61, 122–129.

World Health Organization (WHO), 1997. Anti-tuberculosis Drug Resistance in the World: The WHO/IUATLD Global Project on Antituberculous Drug Resistance Surveillance 1994–97 (Geneva).

World Health Organization (WHO), 2000. WHO/IUATLD Global Project on Anti-tuberculous Drug Resistance Surveillance: Antituberculous Drug Resistance in the World Report 2: Prevalence and Trends. Geneva.

World Health Organization (WHO), 2007. Revised BCG vaccination guidelines for infants at risk for HIV infection. Wkly. Epidemiol. Rec. 82, 193–196.

World Health Organization (WHO), 2008. Who-iutald Global Project on Anti-tuberculosis Drug Resistance Surveillance Anti-tuberculosis drug resistance in the world (report no. 4).

World Health Organization (WHO), 2013. Countdown to 2015: Global Tuberculosis Report 2013 Supplement (Geneva, Switzerland).

World Health Organization (WHO), 2014. Global Tuberculosis Reports.

Yamamoto, T., Nishiura, M., Harada, N., et al., 1958. Electron microscopy of ultra-thin sections of lepra cells and *Mycobacterium leprae*. J. Lepr. 26, 1–8.

Yang, Y., Auguin, D., Delbecq, S., et al., 2011. Structure of the *Mycobacterium tuberculosis* OmpATb protein: a model of an oligomeric channel in the mycobacterial cell wall. Proteins 79, 645–661.

Young, M., Mukamolova, G.V., Kaprelyants, A.S., 2005. Mycobacterial dormancy and its relation to persistence. In: Parish, T. (Ed.), Mycobacterial Molecular Biology. Horizon Scientific Press, Norwich, pp. 265–320.

Zhang, Y., Heym, B., Allen, B., et al., 1992. The catalase-peroxidase gene and isoniazid resistance in tuberculosis. Nature 358, 591–593.

Zhang, Y., Yang, Y., Woods, A., et al., 2001. Resuscitation of dormant *Mycobacterium tuberculosis* by phospholipids or specific peptides. Biochem. Biophys. Res. Commun. 284, 542–547.

Ziehl, F., 1882. Zur Farbung des Tuberkelbacillus 8, 451.
Zopf, W., 1883. Die Spaltpilze. Edward Trewendt, Breslau.
Zumárraga, M., Araujo, F., Cataldi, A., 2014. Population structure genotypes and pathogenic phenotype of *Mycobacterium bovis* in Argentina (and Latin America). Int. J. Mycobacteriol. 4, 30–31.

Index

'*Note*: Page numbers followed by "f" indicate figures and "t" indicate tables.'

FSC
www.fsc.org
MIX
Papier | Fördert
gute Waldnutzung
FSC® C083411

Kolibri 360 GmbH
Ferdinand-Jühlke-Straße 7
99095 Erfurt, Deutschland
produktsicherheit@kolibri360.de